T0191863

Foundations of Hardware IP Protection

Lilian Bossuet · Lionel Torres
Editors

Foundations of Hardware IP Protection

Editors
Lilian Bossuet
Laboratoire Hubert Curien
Jean Monnet University
Saint-Étienne
France

Lionel Torres
Laboratoire LIRMM
Université de Montpellier 2
Montpellier
France

ISBN 978-3-319-84385-8 ISBN 978-3-319-50380-6 (eBook)
DOI 10.1007/978-3-319-50380-6

Softcover reprint of the hardcover 1st edition 2017

Printed on acid-free paper

This Springer imprint is published by Springer Nature
The registered company is Springer International Publishing AG
The registered company address is: Gewerbestrasse 11, 6330 Cham, Switzerland

Preface

The increasing production costs of electronic devices and changes in the design methods of integrated circuits has led to emerging threats in the microelectronics industry such as counterfeiting, illegal copying, reverse engineering and theft. The designing process of a microelectronic Very Large Scale Integration (VLSI) circuit evolved, in last decades, towards a continuously growing « design-reuse » method trend. It is structured today with standard functional blocks vendors (around 50 companies worldwide) delivering « Intellectual Property » blocks, i.e. « IPs ». Those companies face a strong counterfeiting as many other industrial domains, with a strong impact on their business model without any technical solutions to exactly count the dissemination of their models in terms of physical unit devices. Since more than a decade, IP protection has become a critical issue for the microelectronic industry. Electronic devices are increasingly becoming the target of counterfeiting, cloning, illegal copy, theft and malicious hardware insertion (such as hardware Trojans). All these threats cost a lot of money and time to the legal industry. For example in 2014, electronic items counterfeiting was estimated to account for about 7% of the semiconductor market, which represents a loss of around US$ 22 billion in 2014 for the lawful semiconductor industry. Moreover, these threats' impacts are huge employment loss and customer dissatisfaction. However, unlike for software in computer science, protection of hardware IP is not fully included in electrical engineering curriculum. Most of the VLSI designers are not aware about the threats and the means of protection. This book aims to fill the gap by highlighting promising works that attempt to meet the IP protection challenge.

The electronic industry needs solutions to fight against theft, illegal cloning and reverse engineering of intellectual properties. More precisely, designers need salutary hardware, i.e. embedded hardware systems, hardly detectable/difficult to circumvent, inserted in an integrated circuit or a virtual component, used to provide intellectual property information (e.g. watermarking or hardware licensing) and/or to remotely activate the circuit or IP after being manufactured and during its use. When discussing about IP protection, the Digital Right Management (DRM) concept is certainly an important issue for the IP market. The Digital Rights Management (DRM) principle is generally well known for the exchange of files (music, video, etc.), or software management. Specialized solutions concerning professional software are behind a business called "Software License Management". The concept of DRM can be transposed to the IP world, which is a really new concept on this area.

We hope that the readers of this book will learn how an IP can be threatened and to increase the security of the IP by using several different means (hardware obfuscation/camouflaging, watermarking, fingerprinting (PUF), functional locking, remote activation, hardware Trojan detection, protection against hardware Trojan, use of secure element, digital right management, ultralightweight cryptography, etc.). This book will not be like a cookbook as each IP needs specific protection scheme; but it will be like a reference book for design space exploration of security means of IP protection.

Saint-Étienne, France Lilian Bossuet
Montpellier, France Lionel Torres

Contents

Chapter 1
Digital Right Management for IP Protection

Lionel Torres, Pascal Benoit, Jérome Rampon, Renaud Perillat, Dominic Spring, Gael Paul, Stephane Bonniol and Lilian Bossuet

1.1 Introduction

The digital rights management (DRM) is mostly known for the exchange of media files and proprietary software. Specific solutions (e.g., flexlm [1], safenet [2], rlm [3]) are available for professional software, and are at the source of a sector called software license management (SLM). The principle of DRM is well known for managing audio or video file types for access to music or movies in particular around the issues of illegal downloading (excessive infringement to the early 2000s). The methods that apply (secure platform type "iTunes", package deals of type "deezer") does not counteract the primary need that is addressed in the IP authentication, circuits, and fight against counterfeiting. Similarly, in the software field, the business model of software vendors imposes to control their income. Said SLM protection systems have established for 30 years in this sector. Many people are familiar to write software license codes obtained from the provider to switch modes from demonstration on 30 days to a permanent mode. The most common and accessible example is probably the Microsoft Office suite. What we propose to develop is close to SLM concept, in taking the principle of master-slave type used by SLM solutions. It should be noted that marking solutions (identification, bar code, RFID) are available on the market, however the advantage of our approach is to be much more integrated (and indeed secure). The concept of DRM

L. Torres (✉) · P. Benoit
LIRMM, UMR CNRS, University of Montpellier, Montpellier, France
e-mail: Lionel.Torres@lirmm.fr

J. Rampon · R. Perillat · D. Spring · G. Paul · S. Bonniol
Algodone Company, Montpellier, France

L. Bossuet
Laboratoire Hubert Curien, CNRS UMR 5516, Université Jean Monnet, 42000 Saint-Etienne, France

L. Bossuet and L. Torres (eds.), *Foundations of Hardware IP Protection*,
DOI 10.1007/978-3-319-50380-6_1

can also be applied to other fields, and in particular to the one of hardware design. While the DRM concept is accepted and widely used in the software area, this is not the case for the hardware. Indeed, there are no industrial solutions proposed so far, and this becomes an international concern to protect these digital rights, like the recent DARPA projects to fight against counterfeiting and improve the traceability of integrated circuits [4].

Chip counterfeiting is less mediatized than other problems; nevertheless, its consequences can be fatal when we look at the number of electronic devices in a connected hospital or objects related to human. For instance, between 2006 and 2010, the US retailer Vision-Tech circuits delivered almost 60,000 counterfeit circuits to its clients, including the US Navy, Raytheon Missile System [5]. Since then, many cases of counterfeit circuits used in sensitive applications have been reported in the US (military equipment) [6] and are increasingly relayed in the press [7]. The problem of counterfeit integrated circuits has increased significantly in recent years. For example, the number of electronic circuits counterfeits seized by US Customs from 2001 to 2011 has been approximately multiplied by 700 [7]. Between 2007 and 2010, US Customs seized 5.6 million counterfeit electronic products [8]. The estimate of counterfeiting is a minimum of 7 % of the semi-conductor market [9], which represented a loss of about $ 22 billion in 2014 for the legal industry. It is therefore crucial and strategic to implement research projects to protect the intellectual property of IC designers.

In 2013, about 2,000 new ASIC circuits of projects were performed, with high design costs, and nearly 100,000 projects with FPGA circuits to lower design costs, were initiated [10]. These circuits are found in many consumer electronics (smart-phones, tablets, laptops, game consoles, connected objects, etc.) and even safety-critical systems (cars, avionics, military, space, nuclear). For 15 years, nearly all of these new circuits use predefined blocks called Intellectual Property ("IPs") because of their numerous benefits (proven functionality, compatibility, performance, time-to-market, cost and patent enforcement). The design-reuse trend is growing steadily and now it is possible to find more than 100 IP-based elements in a new circuit (e.g., memory, processors, peripherals, communication protocols). The share of IP components represents an average percentage around 70 % of the circuit, up to 90 % in some cases. In order to make this market growing while guaranteeing a return on investment to developers of companies that supply IPs, it is essential to provide an automated control system of IP rights. It is possible to estimate the amount of counterfeited integrated circuits, mainly due to seize from customs services. However, for the IP market it is not possible to observe directly the inner parts of the hardware, since IPs are virtual components when looking at their physical implementations. However, it is clear that IP vendors are threatened in the same way as the circuit provider and undoubtedly they are even more at risk because the copy of an IP is much simpler than a circuit. The key point under the IP activation will be tied to the ability to ensure the uniqueness of the considered component. In the following sections, a dedicated IP (smart lock IP) is discussed, this one based on a Master-Slave protocol brings new features to activate IP features.

1.2 Smart Lock DRM IP Principle

We propose a solution allowing to an IP vendor to take under control the number of sold IPs. Indeed, since no DRM solutions for hardware exist yet, IP Vendors do not have any feedback on how the IP is really used by a customer. The latter can then produce an unlimited number of chips without acknowledging the IP Vendors on the actual number of integrated circuits that are deployed in the market.

As described in Fig. 1.1, currently on the design chain only legal agreements are considered between the IP vendor, IP customer and in some rare cases with the manufacture. The main problem is that the IP vendor has only a partial view of the use of his IP into the final system. To protect its IPs, IP vendors simply need to insert a smart lock IP, which is fully compatible with traditional digital design and verification methodologies.

The main principle of the smart lock IP is the following. Once the chip is manufactured (or the FPGA programmed), each distinct physical instance (each IC devices of FPGA devices) requires a unique runtime license key to activate the functions protected with the smart lock IP. Even though an IP customer may produce thousands or millions of identical devices, each one of them actually requires a unique runtime license key for its activation. The main advantage of our solution is its simplicity: design once, and activate every single IP use (before activation the IP is in standby mode waiting the activation code).

Considering the similar IP integration as presented in Figs. 1.1 and 1.2 proposes two major modifications. The first one is at the IP level where an activation code is inserted. This activation code allows to activate or not the IP. This activation mechanism could be done by inserting logic at the IP level (enabling flip-flop, combinational path control, XOR gates, etc. …). This activation code is included into the activation license. The second principle is based on a primitive that is used in the smart lock IP: a physically unclonable function (PUF). The PUF will serve as

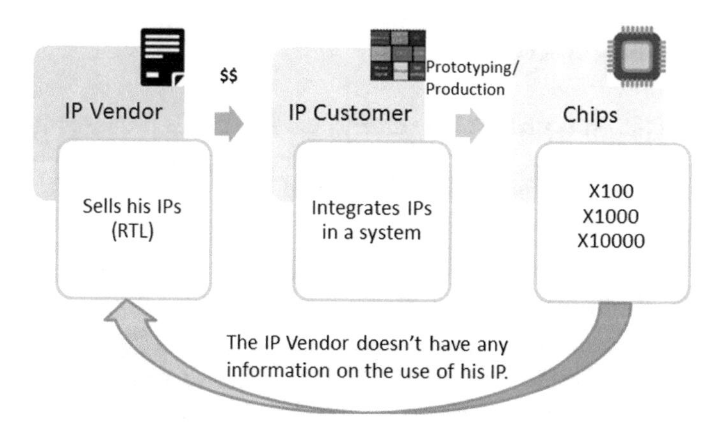

Fig. 1.1 Classical IP flow integration

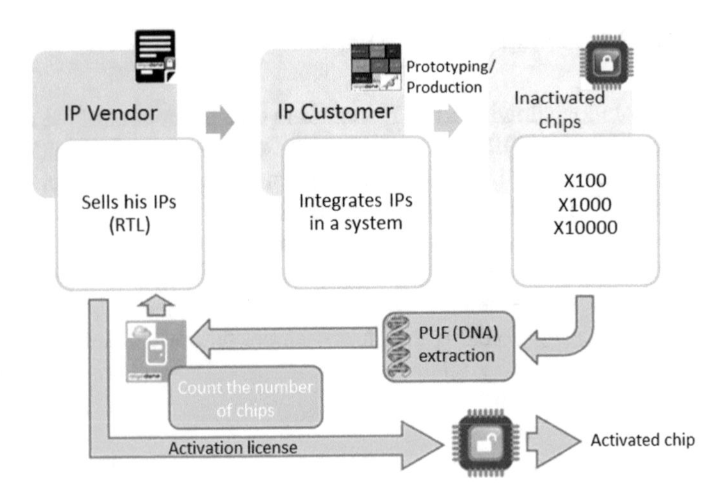

Fig. 1.2 Smart lock IP system principle

a unique device identifier. A PUF is a physical object that can take inputs and generate unpredictable outputs; it is unclonable in that the input/output behavior of a physical copy of one PUF will differ from that of the original one due to some uncontrollable randomness in the process variations. Many literatures are available on PUF, but also some commercial products are now available, as PUF proposed, for instance, by [11, 12] or [13]. After device manufacturing (or during the manufacturing process) the PUF is extracted allowing to have a unique signature per chip. This PUF is used to generate the license (we consider herein that the license is the ciphered stream including activation code of the IP, this license is authenticated and certified on chip) and this license will activate or not the IP. An on-chip license verification block is used to read the identifier associated with the device.

1.3 State of the Art

This state of the art is divided in two parts, the first part is dedicated to the PUF architecture, which is the basis for identification process for DRM flow and the second section is about fingerprint solutions and IP protection.

1.3.1 PUF Principles

The physical unclonable function (PUF) is a hardware security primitive that exploits some physical randomness, introduced explicitly or indirectly into a device, to generate device unique information used to address security-related

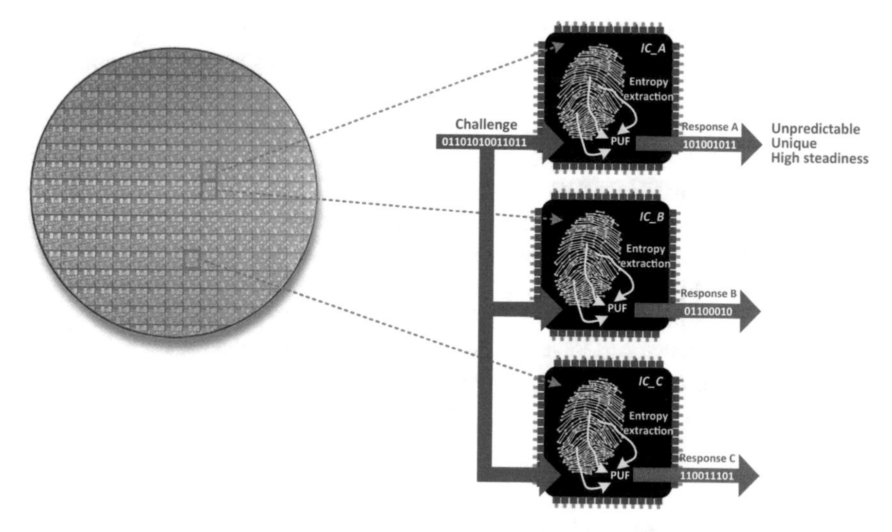

Fig. 1.3 Illustration of the behavior of PUF

problems. Basically, a PUF is a function that one-way maps a set of inputs, namely challenges, to a set of outputs, defined as responses, forming a set of challenge-response pairs (CRPs) that are unique for each device where the PUF has been implemented. Moreover, PUFs should be unclonable and tamper evident, meaning that it is unfeasible for an attacker to build another PUF providing the same original CRPs set, and that invasive attacks destroy the PUF, as they can be easily detected. Furthermore, PUF responses should be persistent and unpredictable, and it should not be possible to discover the applied challenge given its output response. Figure 1.3 is an illustration of the behavior of a silicon PUF when three chips from the same wafer embed the same PUF, they are packaged in three integrated circuits, and after receiving the same challenge, each PUF provides a unique, unpredictable, and steady response by performing an extraction of the entropy which comes from variations in the CMOS process.

Among all types of PUFs, this section focuses on silicon PUFs, introduced [14], exploiting variations in integrated circuit manufacturing, inherently random across different dies and wafers, to generate robust, unclonable, unpredictable, and chip-specific outputs. Due to their underlying mechanism, silicon PUFs do not require any alteration of manufacturing processes, and several designs are available for both field-programmable gate arrays (FPGAs) and application-specific integrated circuits (ASICs).

Since exploited properties are electrical, the responses are inherently affected by noise. The environmental and working conditions, such as the temperature and the supplied voltage, can alter PUFs responses. For this reason, often the PUF is followed by a code corrector block to guarantee the uniqueness of the response. According to different physical sources of imperfections, silicon PUFs can be categorized in delay-based PUFs and memory-based PUFs. While the former requires additional hardware resources, because it involves a time measurement, the latter is based on the random power-up values of storage elements, inherently present in practically any Integrated.

The most discussed memory-based PUF are the SRAM PUF [14] and the D flip-flop PUF [15]. As for the delay-based category, there are the Arbiter PUF [16], the Ring Oscillator (RO) PUF [17–19], the Butterfly PUF [20], and the Anderson PUF [21] for the delay-based family. SRAM PUFs are not suitable for all FPGA families, hence the introduction of D flip-flop PUFs. Arbiter PUFs and Butterfly PUFs require layout symmetry that is hard to satisfy on FPGAs due to the lack of routing control. Conversely, the RO PUF [22] is suitable for every silicon technology, and hence can be easily adopted as PUF technique for both the FPGA and ASIC technology. Indeed, the RO is a primitive easily implementable in terms of design.

1.3.2 Fingerprint and IP Protection

PUF are systems therefore allowing embedded measurement of physical changes in the manufacturing process, they are part of an authentication category by physical characterization commonly called fingerprinting. There are other methods of fingerprinting operating system such as power consumption, such as the method proposed [23] and generalized by [24], or temperature evolution as the method proposed by the company ALGOTRONIX [25]. These two methods are interesting to copy detection but unlike PUF, they do not allow the implementation of a challenge/response protocol (or client/server) to identify and unlock remote IP.

When fingerprinting is not possible, the designer has the possibility to add an hardware detectable tag to differentiate its IP and to detect a copy. This is known as watermarking. Regarding watermarking, state of the art is wide and it was featured in a recent study [26]. One of the difficulties of the methods of watermarking consists of the tag verification; many methods of marking do not propose a simple method to check these tags [24]. Moreover, as indicated, watermarking is a technique for the detection of copy but without preventing it.

In addition to PUF approaches, it is necessary to consider the state of the art of security approaches, including FPGAs made, namely by considering technical security of their configuration file (called bitstream). The aim is mainly to ensure that the bitstream is secure enough to prevent all forms of cloning. Works such as [27–32], have been taken up by manufacturers such as Xilinx, Altera, and Microsemi to propose commercial solutions. However, one cannot really talk about DRM, but most of FPGA bitstream protection techniques.

Furthermore, the company KayaInstrument [33] offers a generic solution and ad hoc to protect against any FPGA bitstream cloning, using an addi-silent component. However, if this solution is interesting, it does not offer the possibility to authenticate IP or activate them. Moreover, it is not possible to use this solution when there are multiple IP. This case is complicated, few studies have addressed in [34] or [35] a solution based on a complex protocol involving a third trusted entity as described into [36]. The work [37] or more recently [38] propose to intervene directly at the FPGA manufacturer.

The work of the Algotronix company for the world of FPGA [39] in 2002 shows that a solution could be based on an asymmetric protocol and a secret key contained in the FPGA. If this work has opened the DRM path in the world of FPGA, the fact remains that the proposed solution is dependent on FPGA vendors, it requires a public key encryption algorithm and this one is complex to design on a minimum area of silicon. As mentioned, lot of works have been proposed for FPGA world, and some have proposed extensions for SoC approach. The DRM flow that we consider is a generic flow for any target (FPGA or SoC), considering a trusted framework from license generator to activation code of the IP. But clearly another aspect not considered in this chapter is the fact that DRM is a way to better monetize IPs and the silicon market [40].

1.4 DRM Flow

An example of the process for protecting and activating an IP block will now be described in more detail (Fig. 1.4).

While this example is based on the protection of an IP block, as mentioned above, in alternative embodiments it could be applied to the protection of other

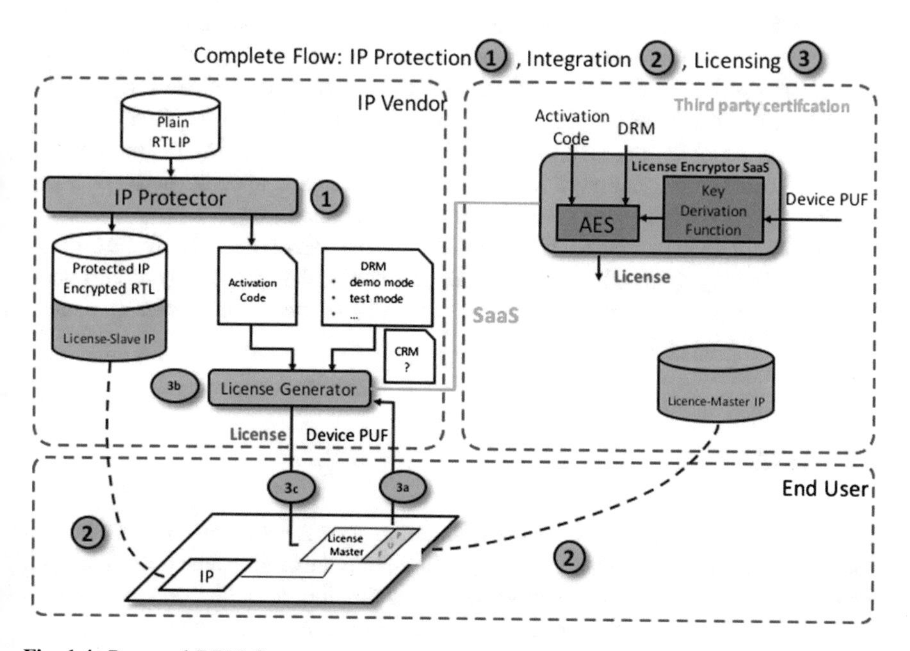

Fig. 1.4 Proposed DRM flow

types of circuits. At the IP vendor, a database (1) for example stores one or more IP blocks represented in a hardware description language such as plain RTL (Register Transfer Level).

An IP protector block (IP PROTECTOR) comprises software for executing an IP protection function, which is used to modify one or more of the IP blocks from the database to render the circuit inactive (refer to activation code described in previous section). In particular, the IP block is altered such that an activation code is required to unlock its functionalities.

The modification of the IP block for example involves inserting, into the hardware description file, one or more logic gates into one or more signal paths of the IP block. These logic gates for example permit the propagation of the unaltered signal along the signal path only if one or more correct activation bits are provided at one or more input nodes of the inserted logic gates. Only the protected IP will be delivered to the designer or the end-user.

In the case that the license is generated by the IP vendor, the activation code needed for unlocking the protected IP block or blocks of the integrated circuit is provided to the license generator. The license generator is used to generate a HW license, which is transmitted to the license server (the smart lock DRM IP) of the end-user device (in fact the license could be stored in some non-volatile memory or any permanent storage depending of the system configuration and accesses).

The license generator receives also the device identifier (PUF) from the final device too. So, the license is essentially based on this identifier (PUF) and on the activation code. In some cases, the license also incorporates other data. These data for example indicates any limit on the duration of activation of an IP block, in the case that the license is a temporary license. It may also indicate a limitation of the license to one or more hardware types, such as a technology of ASIC, FPGA family, or specific features of an IP that are to be unlocked. We can imagine also that IP activation is depending of different sources of sensors: GPS position, temperature, acceleration, and so on, enabling features depending of the environment. Of course the generated license is ciphered and a specific derivation key function (based on the PUF) is used to generate the ciphered key.

Another interesting approach to consider is that rather than the license being generated directly by the IP vendor, the license could be generated by a third party (generally certified) by a web service through a SaaS mode. In all cases the same principle of license generator is used.

1.5 DRM Integration in SoC

At circuit level, the smart lock IP is distributed in several elements into the SoC. As described into Fig. 1.5 the activation logic is directly inserted on each IP to protect. For each IP, by a small integrated wrapper it is possible to define several features: full activation or not, partial activation or not, activation delay, demo mode, and so on. An integrated master-server is able to deliver in secure way a dedicated key for

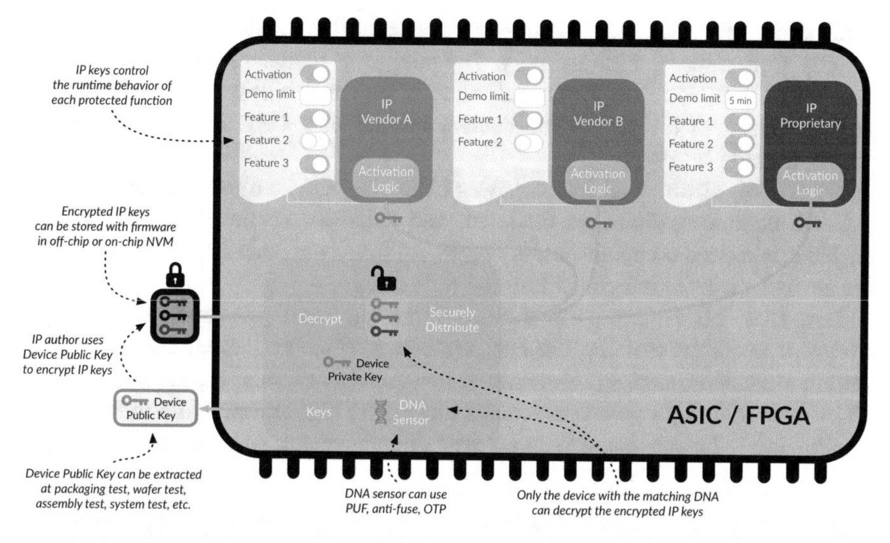

Fig. 1.5 Smart lock IP SoC integration

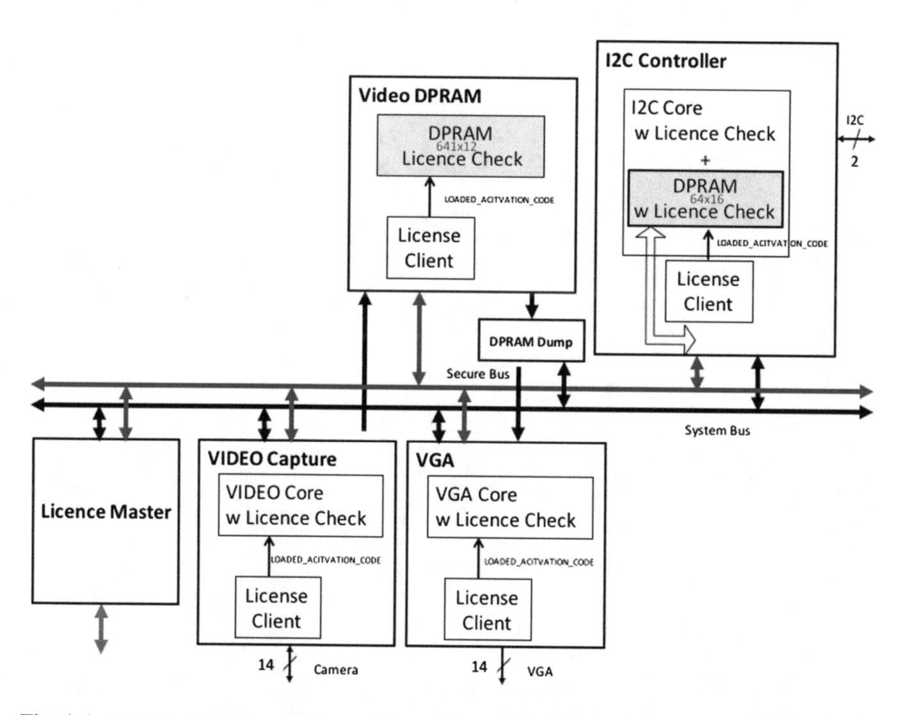

Fig. 1.6 Smart lock IP integration on video codec architecture

each IP to enable these services. Theses keys are on-chip provided by an integrated master-server and transmitted to the IP by a secure link (based on challenge responses mechanism). The integrated trusted server is the entity in charge to receive the HW license. This license may come from the SaaS framework or directly from a secure flash memory. The integrated master-server is also in charge to extract the PUF of the device and the PUF signature will serve as public/private key protocol and to identify the device. Instead of a PUF an OTP or anti-fuse structure could be used.

Of course, all the exchanges between the integrated trusted master-server and external communications are ciphered, and a public/private protocol is used to enable this trusted communication. Figure 1.5 gives the main synoptic on how that the smart lock IP is integrated into the SoC design.

Based on this concept, a first architecture has been proposed to validate the overall concept on real IP, this one proposes a video codec, with a DPRAM IP storing the video frame, a I2C controller for external links and application control, a VGA core able to propose basic image processing algorithms, and a video capture IP allowing to be connected to an external digital camera. As describes on Fig. 1.6,

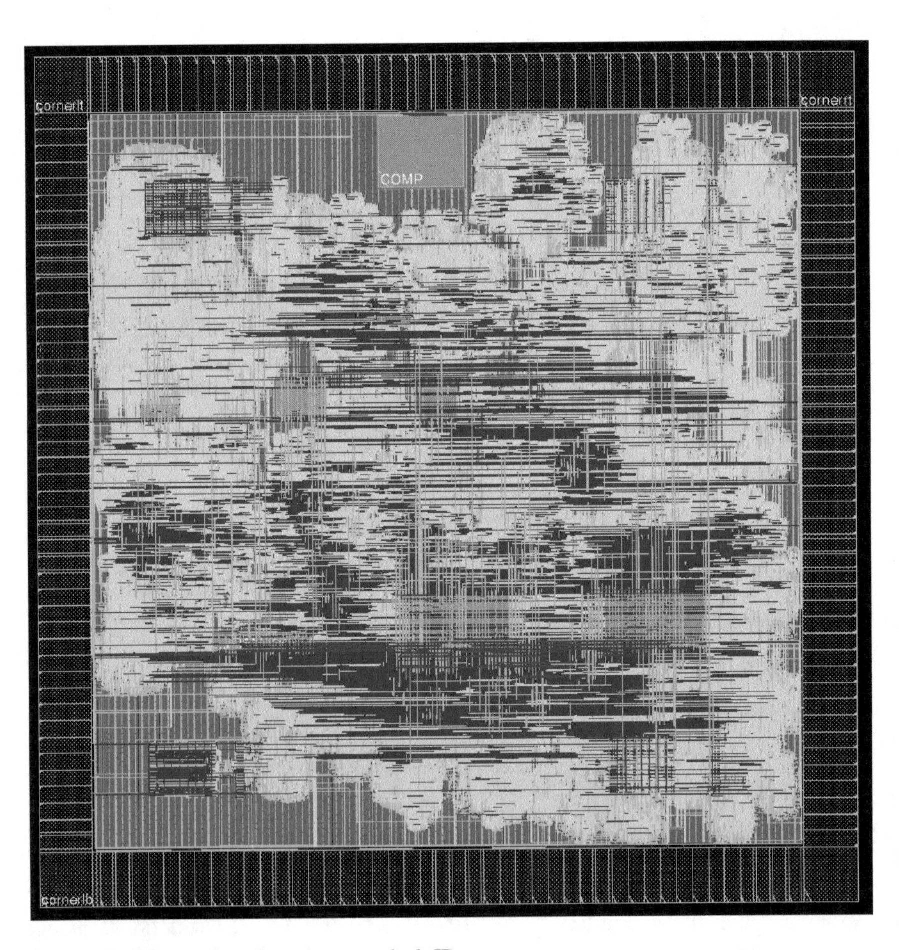

Fig. 1.7 SoC layout based on the smart lock IP

for each IP a license client is added to each IP permitting the IP activation under the control of the trusted license server. A secure bus is also proposed to connect the trusted license server to the client server of each IP. The protocol is based on a challenge–response authentication. As mentioned in previous section, license server extracts the PUF (or the device ID) and receives and decodes the ciphered license providing specific activation keys for each IP.

A full implementation in 65 nm CMOS technology is proposed to validate the overall concept. Figure 1.7 shows the actual SoC layout. The overall SoC is about 1 Million of gates for 6mm2 silicon fingerprint. The trusted license server itself is about 60Kgates (6 % of the overall design).

1.6 Conclusion

We proposed in this chapter, a technology able to manage hardware digital rights (DRM) in a very similar way to the principles of software licensing solutions. This solution could fix strong issues of the semiconductor market: recurrent revenue in the IC world, leading business models not available in IC world, i.e., “in-app” purchase mode, Upgrade, Upsell, Try-before-you-buy, Pay-per-Use, and so on.

We show also how this DRM solution could really be implemented as an IP for SoC or FPGA platform. This IP is based on a DRM controller, a dedicated wrapper for IP activation, and a chip identity can possibly be implemented with a physical unclonable function (PUF) delivering a unique ID for a given manufactured chip.

The smart lock IP delivers features to build an on-chip hardware infrastructure enabling a secure licensing of hardware component instances, per manufactured chip.

Three components are necessary

- The IP Activator is used to instrument components for feature-based activation;
- The Sensor Authenticator is used to deliver a safe chip identity mechanism. It is the reference for the license generation;
- The DRM Controller to extract the chip identity for license request and then to read the license and activate the instrumented hardware elements (IP/IC).

The approach described is a hardware license generator working in SaaS mode (during the manufacturing process or not). It includes some Silicon Management System based on a Hardware Development Kit and a Software Suite to generate the Hardware license. Multiple instrumented IPs from different vendors at different levels of hierarchy can be connected with one DRM Controller and one authenticated Sensor, through a specific bus. At design phase, designer simply needs to insert the DRM controller, which is fully compatible with traditional digital design and verification methodologies. Once the chip is manufactured or the FPGA programmed, each distinct physical instance requires a unique runtime license key to activate the functions protected with a secured license protocol. Even though thousands or millions of identical devices are produced, each one of them requires a unique license key to control its activation at operational runtime. The objective of this work was really to demonstrate that DRM is now a real concept, easy to include in new SoC platform and bringing new facilities for the end-user application. Finally, the proposed solution is fully compatible with the actual design flow and manufacturing process avoiding additional constraints.

References

1. FlexLM, Solution for applications producers, http://www.flexerasoftware.com/
2. SafeNet, World leading data protection and software monetization, http://www.safenet-inc.com/
3. RLM, Reprise license manager, http://www.reprisesoftware.com
4. DARPA, http://www.darpa.mil/news-events/2014-02-24
5. A.Z.P.E. Chaudhry, Protecting Your Intellectual Property Rights, The Global Growth of Counterfeit Trade (2013)
6. S. Maynard, Trusted foundry be safe. be sure. be trusted. Trusted Manufacturing of Integrated Circuits for the Department of Defenses (2010), http://www.trustedfoundryprogram.org
7. C. Gorman, Counterfeit chips on the rise (2012)
8. AGMA, Alliance for Gray Markets and Counterfeit Adatement, http://www.agmaglobal.org
9. M. Pecht, S. Tiku, Bogus! Electronic manufacturing and consumers confront a rising tide of counterfeit electronics. IEEE Spectrum (2006)
10. Gartner, http://www.gartner.com
11. https://www.intrinsic-id.com
12. http://www.ictk.com/servicenproduct/puf
13. http://secure-ic.com/sic-trusted-puf
14. B. Gassend, D. Clarke, M. Van Dijk, S. Devadas, Silicon physical random functions, in *Proceedings of the 9th ACM Conference on Computer and Communications Security* (ACM, 2002), pp. 148–160
15. D.E. Holcomb, W.P. Burleson, K. Fu, Power-up SRAM state as an identifying fingerprint and source of true random numbers. IEEE Trans. Comput. **58**(9), 1198–1210 (2009)
16. V. van der Leest, G.-J. Schrijen, H. Handschuh, P. Tuyls, Hardware intrinsic security from D flip-flops, in *Proceedings of the Fifth ACM Workshop on Scalable Trusted Computing* (ACM, 2010), pp. 53–62
17. D. Lim, J.W. Lee, B. Gassend, G.E. Suh, M. Van Dijk, S. Devadas, Extracting secret keys from integrated circuits. IEEE Trans. Very Large Scale Integr. (VLSI) Syst. 13(10), 1200–1205 (2005)
18. G.E. Suh, S. Devadas, Physical unclonable functions for device authentication and secret key generation, in *Proceedings of the 44th Annual Design Automation Conference* (ACM, 2007), pp. 9–14
19. A. Maiti, J. Casarona, L. McHale, P. Schaumont, A large scale characterization of RO-PUF, in *2010 IEEE International Symposium on Hardware-Oriented Security and Trust (HOST)* (IEEE, 2010), pp. 94–99
20. A. Maiti, P. Schaumont, Improved ring oscillator puf: an fpga friendly secure primitive. J. Cryptol. **24**(2), 375–397 (2011)
21. S.S. Kumar, J. Guajardo, R. Maes, G.-J. Schrijen, P. Tuyls, The butterfly puf protecting ip on every fpga, in *IEEE International Workshop on Hardware-Oriented Security and Trust, 2008. HOST 2008* (IEEE, 2008), pp. 67–70

22. J.H. Anderson, A puf design for secure fpga-based embedded systems, in *Proceedings of the 2010 Asia and South Pacific Design Automation Conference* (IEEE Press, 2010), pp. 1–6
23. M. Barbareschi, G.D. Natale, L. Torres, Ring oscillators analysis for security purposes in Spartan-6 FPGAs. Elsevier Microprocess. Microsyst. doi:10.1016/j.micpro.2016.06.005
24. S. Kerckhof, F. Durvaux, F.X. Standaert, B. Gérard, Intellectual property protection for FPGA designs with soft physical hash functions: first experimental results. HOST, pp. 7–12 (2013)
25. C. Marchand, L. Bossuet, E. Jung, IP watermark verification based on power consumption analysis. SOCC 2014
26. C. Marsh, T. Kean, D. Mclaren, Protecting designs with a passive thermal tag. ICECS, pp. 218–221 (2008)
27. B. Le Gal, L. Bossuet, Automatic low-cost IP watermarking technique based on output mark insertion. J. Des. Autom. Embed Syst. **16**(2), 71–92 (2012). Springer
28. L. Bossuet, G. Gogniat, W. Burleson, Dynamically configurable security for SRAM FPGA bitstreams. Int. J. Embed. Syst. **2**(1/2), 73–85 (2006). Interscience Publishers
29. Y. Hori, A. Satoh, H. Sakane, K. Toda, Bitstream encryption and authentication with AES-GCM in dynamically reconfigurable systems. FPL, pp. 23–28 (2008)
30. S. Drimer, M.G. Kuhn, A Protocol for Secure Remote Updates of FPGA Configurations. ARC, Springer, LNCS, vol. 5453, pp. 50–61 (2009)
31. F. Devic, B. Badrignans, L. Torres, Secure protocol implementation for remote bitstream update preventing replay attacks on FPGAs. FPL, pp. 179–182 (2010)
32. A. Braeken, J. Genoe, S. Kubera, N. Mentens, A. Touhafi, I. Verbauwhede, Y. Verbelen, J. Vliegen, K. Wouters, Secure remote reconfiguration of an FPGA-based embedded system. ReCoSoC, pp. 1–6 (2011)
33. L. Bossuet, V. Fischer, L. Gaspar, L. Torres, G. Gogniat, Disposable configuration of remotely reconfigurable systems. Microprocess. Microsyst. Embed. Hardw. Des. **39**(6), 382–392 (2015). Elsevier
34. http://www.kayainstruments.com
35. S. Drimer, T. Güneysu, M.G. Kuhn, C. Paar, Protecting multiple cores in a single FPGA design (2008), http://www.saardrimer.com/sd410/papers/protect_many_cores.pdf
36. J. Vliegen, D. Koch, N. Mentens, D. Schellekens, I. Verbauwhede, Practical feasibility evaluation and improvement of a pay-per-use licensing scheme for hardware IP cores in Xilinx FPGAs. J
37. E. Simpson, P. Schaumont, Offline hardware/software authentication for reconfigurable platforms. CHES, Springer, LNCS vol. 4249, pp. 311–323 (2006)
38. T. Guneysu, B. Moller, C. Paar, Dynamic intellectual property protection for reconfigurable devices. ICFPT (2007)
39. L. Zhang, C.H. Chang, A pragmatic per-device licensing scheme for hardware ip cores on SRAM-based FPGAs. IEEE Trans. Inf. Forensics Secur. **9**(11), 1893–1905 (2014)
40. T. Kean, Cryptographic rights management of FPGA intellectual property cores, in Proceedings Tenth ACM International Symposium on FPGAs, Monterey CA, 2002 42. Global Semiconductor
41. Global Semiconductor Alliance, http://gsaglobal.org

Chapter 2
Turning Electronic Circuits Features into On-Chip Locks

Brice Colombier, Lilian Bossuet and David Hély

2.1 Introduction and Context

Following Moore's law, electronic systems are increasingly complex and powerful. Their complexity is following a similar trend, forcing designers to adopt a modular approach when designing such systems. Thus, a design-and-reuse approach is followed, in which functional building blocks are put together by system integrators. These blocks are provided by IP cores designers, who must transfer their complete design in order to have it implemented correctly. However, such a situation necessarily leads to abuses, since the designer cannot control the number of instances implemented from its original design. It results in overbuilding IP cores and counterfeiting of integrated circuits, and the trend is growing. Multiple cases have been reported in recent years [1–3].

In order to answer this issue, the circuit can be provided as initially locked. It is then nonfunctional and should be unlocked in order to be used. The unlocking procedure can be initiated only by the designer, allowing precise audit of the number of instances of the protected design. This is referred to as hardware metering [4]. In case the design has been obtained illegally, either from overbuilding or counterfeiting, it remains locked and therefore unusable.

There are several ways to achieve locking of a circuit. Among them, modifying the combinational logic is a way. It is presented in details in Chap. 3.

B. Colombier (✉) · L. Bossuet
Hubert Curien Laboratory, UMR CNRS 5516, University of Lyon, Saint-étienne, France
e-mail: b.colombier@univ-st-etienne.fr

L. Bossuet
e-mail: lilian.bossuet@univ-st-etienne.fr

D. Hély
LCIS, Grenoble Institute of Technology, Valence, France
e-mail: david.hely@lcis.grenoble-inp.fr

L. Bossuet and L. Torres (eds.), *Foundations of Hardware IP Protection*,
DOI 10.1007/978-3-319-50380-6_2

Another interesting approach is to look into hardware Trojans [5] that target Denial-Of-Service attacks. Such attacks are closely related to the kind of remote locking we want to achieve here. Therefore, the actual means that are used by hardware Trojans to trigger and achieve a Denial-Of-Service attack could be turned into remote locking techniques. It effectively turns malicious hardware into salutary hardware [6]. In order to be worth considering from the designer's perspective, a design protection scheme must also be cheap in terms of additional hardware resources required to implement it. Indeed, if the economic losses associated with the illegal actions are actually less expensive than the protection scheme itself, the latter becomes unsuitable. Therefore, the protection scheme should be as lightweight as possible, and occupy a very small area on the protected chip. This characteristic is very common for hardware Trojans, which are usually found in the form of tiny and stealthy modifications of the original design.

The point here is to achieve locking by targeting very sensitive components. These components should be crucial to the proper functioning of the system. Thus, their disablement will render the system absolutely unusable, indeed achieving locking of the whole design. These points of action can be thought of as *single points of failure*, which correct behaviour is absolutely necessary to the overall system.

In this chapter, we show that such points can be found in the vast majority of complex electronic systems [7]. This is valuable since the design protection scheme should be usable for any type of design and should not depend on specific design features.

This chapter is organised as follows. Section 2.2 identifies the features which could be turned into on-chip locks and provides a comparison of them using several criteria. Section 2.3 shows how such features can be modified to disturb the circuit's operation. Section 2.4 gives implementation results on FPGA. Two reference designs and two FPGA families were used. Section 2.5 proposes a discussion on partial locking, which is an interesting way to provide a circuit in evaluation mode.

2.2 Features Usable as Locking Means

Figure 2.1 shows a complex electronic system. We highlighted the following common features which can be turned into on-chip locks:

1. The clock circuitry,
2. The inputs/outputs,
3. The processor,
4. The interconnection buses,
5. The system controller,
6. The analogue components.

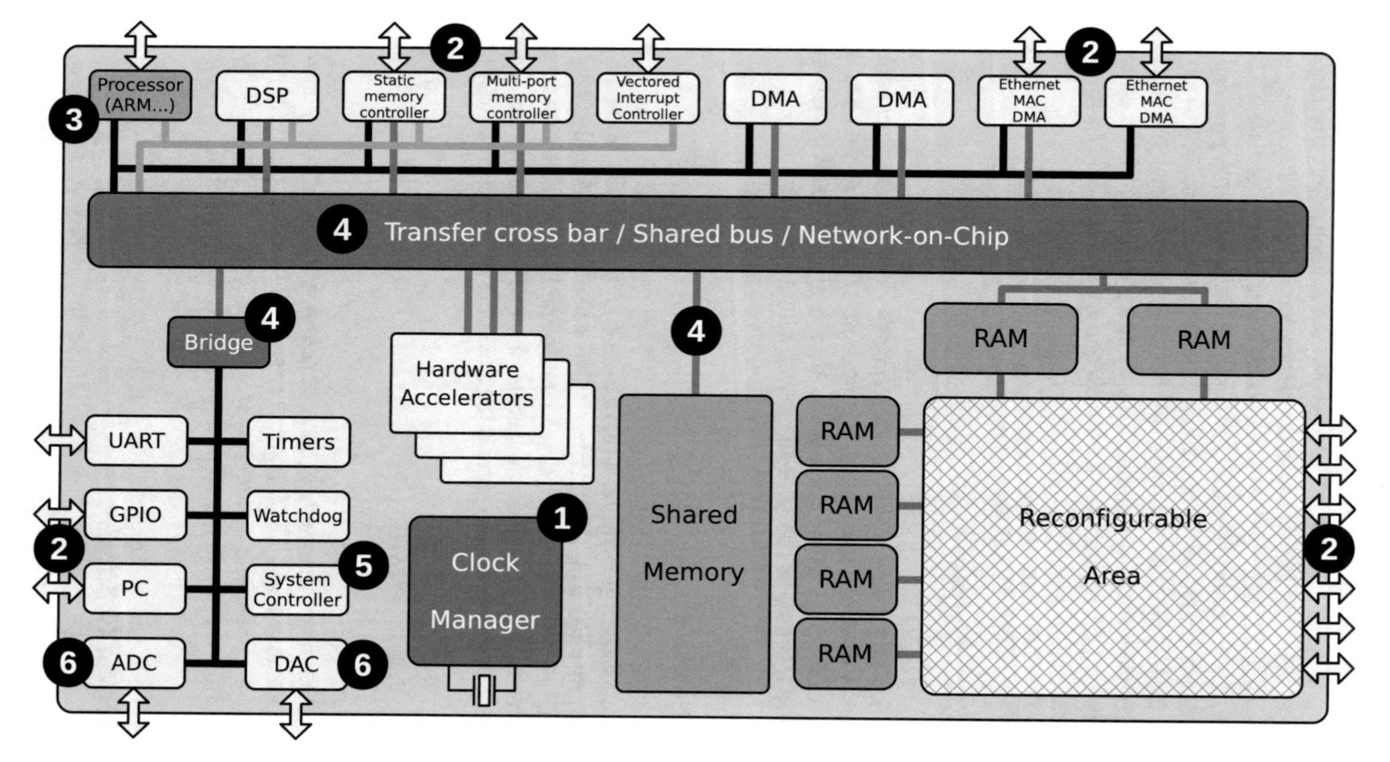

Fig. 2.1 SoC features which can be turned into on-chip locks

2.2.1 Clock Circuitry

The first feature that is immediately identifiable as a locking point is the clock circuit. Indeed, it is a universal feature found in most digital designs. Moreover, good operation of the circuit is heavily dependent on the clock signal. Thus by acting on the clock signal, it is possible to disable the circuit, making it effectively locked. Another interesting characteristic of the clock is that its frequency is related to the device's performances. Hence by dynamically shifting the clock frequency, it is possible to alter the performances of the circuit. This could be used to provide an evaluation version of the device, operating at a lower frequency and exhibiting a lower level of performance.

2.2.2 Inputs/Outputs

All electronic designs have input and output ports to interact with other components. By temporarily disabling these ports, it is possible to prevent new data to be sent to the design. Even though it does not make the design unusable itself, it makes it almost useless, since it is then not possible to interact with it anymore.

2.2.3 Processor

When a processor is present in a digital design, it is usually a central component. Such processor can be either hardwired or soft-core. A soft-core processor is described in a hardware description language and implemented in reconfigurable resources. In essence, the processor executes a sequence of instruction. One way to alter its functionality is then to prevent the execution of new instructions.

2.2.4 Buses

Interconnection buses are the backbone of complex systems. They allow multiple IP cores to communicate. The integrity of the information exchanged between the different sub-modules of a system is a crucial requirement. Therefore, by altering this information, it is possible to render the system nonfunctional.

2.2.5 System Controller

The control logic of complex designs is usually handled by an FSM. By modifying this FSM's states, it is possible to alter the operation of the circuit. Another possibility is to add extra states to control access to the normal mode of operation.

2.2.6 Analogue Components

In order to handle physical data, a design can integrate analogue components. Such components are precisely calibrated to suit the needs of the designer. By altering this calibration, their behaviour can be altered.

Another important analogue component of the design is the power supply module. By shutting down specific areas of the design, they can be efficiently disabled. This feature is called *power gating* and is already implemented in some designs to reduce power consumption.

2.2.7 Global Comparison

After identifying these features, we can have a first overview of their pros and cons. Table 2.1 presents a qualitative comparison.

The first criterion used to evaluate the features is the impact on performance. It describes how the performance of the circuit is affected during normal operation. Modifying the clock circuitry has low impact on the performance, although the clock characteristics such as the jitter can be affected if the modification is poorly handled. Acting on the inputs/outputs, the FSM or the processor does not have any impact on the circuit's performance. Conversely, modifying the buses can lead to slower data rates and increase the latency. Similarly, modifying the calibration of analogue components can reduce their efficiency.

The second criterion is the ease of locking/unlocking. It quantifies how simple it is to implement locking using the corresponding feature. It also shows how easy it is to fall back into normal behaviour after an unlocking request has been received. For example, acting on the clock circuitry or the inputs is simple. They can be easily disabled and enabled again. On the other hand, modifying the processor to be able

Table 2.1 Qualitative comparison of the presented features when being used as on-chip locks

Feature modified for locking	Evaluation criterion				
	Impact of the locking scheme on performance	Ease of dynamic (un)locking	Efficiency/impact on functionnality	Partial locking	Overall suitability
Clock	Low	High	High	yes	• • •
Inputs/outputs	None	High	Medium	no	• • ○
Processor	None	Medium	High	no	• • ○
Buses	Medium	Medium	High	no	•○○
FSM	None	Low	High	no	•○○
Analogue parts	Medium	Low	Medium	yes	•○○

to stop it can be complicated. Similarly, tampering with the buses can lead to unexpected behaviour. In both cases, correctly coming back to normal behaviour might not be guaranteed. When modifying the FSM, locking refers to entering "hidden" states, corresponding to altered operation. Therefore, locking or unlocking requires to have access to the FSM inputs. This is not guaranteed, and most designs do not allow to transition between FSM states so easily. Similarly, modifying the calibration of analogue components can be hard to achieve.

When modifying a feature to achieve locking, the impact on the circuit functionality should be as high as possible, to make is completely unusable. Disabling the clock, processor, buses or entering "hidden" FSM states systematically leads to complete locking. On the other hand disabling the inputs/outputs or altering the characteristics of analogue components has medium impact, which will depend on the usage.

Finally, partial locking is possible with some of the described features. We define partial locking as a state in which the design has a correct behaviour, but has a lower level of performance. This is achievable only by acting on the clock or the analogue components. Modifying the clock frequency directly affects the designs performance. Likewise, altering the calibration of analogue components can make them perform poorly.

The final column on the right of Table 2.1 gives an overall suitability estimation for the feature. It estimates how suited the feature is in order to be turned into an on-chip lock.

2.3 Practical Transformation into On-Chip Locks

We now give means how to turn the features presented in Sect. 2.2 into on-chip locks. Analogue components modification is not discussed here.

2.3.1 Clock Circuitry

In practise, acting on the clock circuitry can be achieved in two ways. The first one is to use a modified clock-gating module. The second one makes use of the reconfiguration capabilities of some phase-locked loops (PLLs).

2.3.1.1 Clock Gating

An already approved method to act on the clock is clock gating. It is commonly used to reduce power consumption by not clocking the unused regions of the circuit. Therefore, it could also be used to make the circuit unusable.

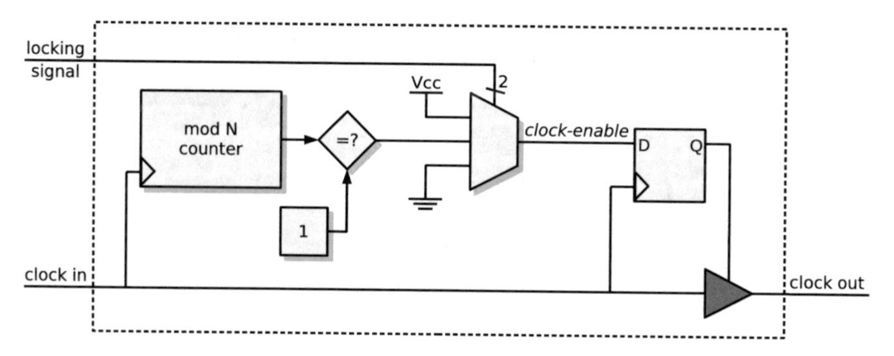

Fig. 2.2 Clock-gating module acting on a clock buffer (in *dark-grey*)

To achieve clock gating, we insert a specific module on the clock signal path. This module is shown in Fig. 2.2. It does not require to add extra logic on the clock signal path itself, but rather makes use of the clock-enable inputs of existing clock buffers.

Here, the clock-enable input can be driven by three different signals. The first one, which corresponds to a high logic level (V_{cc}), allows to leave the clock signal unchanged. In this case, the circuit is totally unlocked. The second one is the result of the comparison between the output of an n-bit counter and the n-bit value 1. Thus, the clock buffer is only active when the counter is equal to 1. In practical terms, the output frequency is then divided by 2^n, where n is the size of the counter.

In this case, the output clock does not have a 50 % duty cycle. Instead, the duty cycle α obtained from the division is given in Eq. 2.1.

$$\alpha = \frac{t_H}{T} = \frac{t/2}{n.t} = \frac{1}{2n} \tag{2.1}$$

In such case, if the frequency is chosen to be divided by a large number, the duty cycle can drop to low values. However, if the setup times were not violated with the original frequency, then they will not be either with the divided frequency. The waveforms obtained from the three different modes presented here are shown in Fig. 2.3: Fig. 2.3a shows the original clock and Fig. 2.3b shows the divided clock. Here, the division factor is 2. We see that the duty cycle is not 50 % but 25 %. This can be useful to provide the design in "evaluation" mode. The operating frequency is twice lower, and so is the performance. Finally, Fig. 2.3c shows the gated clock. In this mode, the design does not function at all.

Acting on the clock has multiple advantages. First of all, it is a powerful way to completely disable the circuit. If the clock is not provided, most of the circuit's elements do not operate. Then, it requires very few additional logic resources. In the module presented in Fig. 2.2, only one counter, one comparator, one multiplexer and one D flip-flop are used. It also allows to reduce the operating frequency, effectively getting the circuit to operate in evaluation mode.

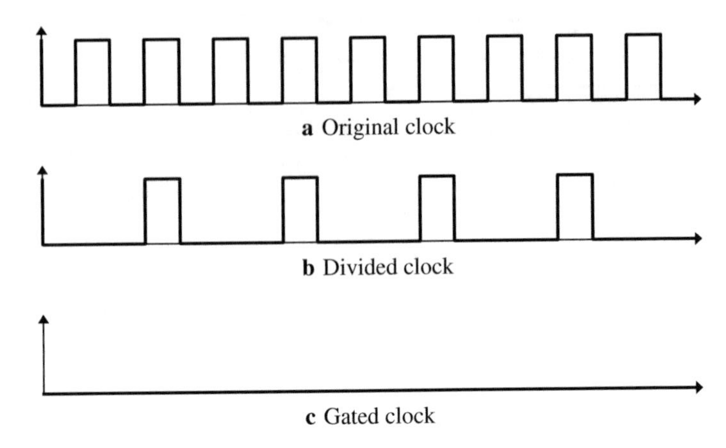

Fig. 2.3 **a** Original clock, **b** divided clock **c** gated clock

The main drawback of such a scheme is that it inherently requires to alter the clock distribution network. This can be problematic in certain designs because the clock distribution network is usually very precisely tuned to meet timing requirements.

Therefore, next subsection proposes another way to act on the clock signal, by dynamically modifying the PLL configuration.

2.3.1.2 Dynamic Phase-Locked Loop (PLL) Reconfiguration

Another way to act on the clock signal is to directly deal with the phase-locked loop (PLL). In most of the integrated circuits, the clock signal is handled by a PLL. It allows to generate multiple clean clock signals, which can have a different frequency, to different parts of the circuit. It is then distributed by the clock tree.

In modern FPGAs, such as Altera Arria V, Cyclone V or Stratix V families [8], the PLL can be dynamically reconfigured. That is, the multiplication and division factors can be dynamically tuned. The PLL is then actually used as a frequency synthesiser.

The output frequency of the PLL is given by the following formula:

$$f_{out} = f_{in} \cdot \frac{M}{N.C} \tag{2.2}$$

C is the post-scale output counter. M is the feedback counter. N is the prescale counter. The values for M, N and C can be dynamically changed in order to tune the operating frequency.

In order to do this on FPGAs, a vendor-specific IP core must be instantiated. It is controlled by a dedicated finite state machine (FSM), responsible for providing the C, M and N values. Once the parameters have been sent, the actual reconfiguration starts. The PLL unlocks, and then locks again on the new frequency. This is illustrated in Fig. 2.4.

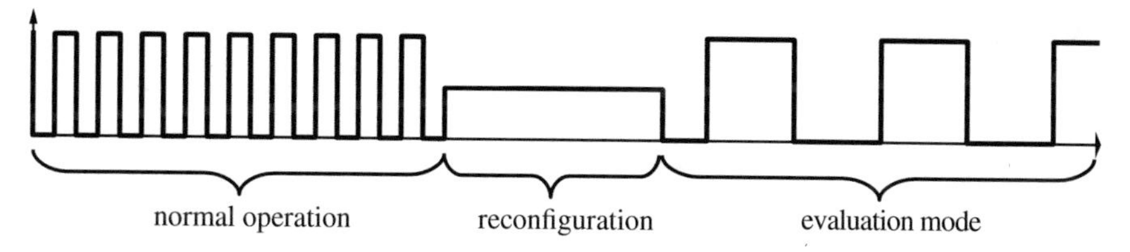

Fig. 2.4 Output clock during a reconfiguration

The normal mode of operation corresponds to the maximum frequency. Then the PLL is reconfigured, and locks again to the new frequency. This one is lower and corresponds to the circuit operating in evaluation mode.

The main advantage of using the PLL as a frequency synthesiser is the flexibility it provides. Indeed, by individually setting the M, N and C parameters, it is possible to precisely tune the frequency. Moreover, since the reconfiguration process is natively supported by the PLL, the designer ensures that the output clock meets the specifications.

On the other hand, such reconfiguration feature might not be found in all the PLLs. For example, only recent versions of Altera's FPGAs support this feature. Another drawback is the area overhead. Indeed, instantiating the reconfiguration engine and the controlling FSM requires a lot of logic resources. This will be extensively discussed in Sect. 2.4.1.

2.3.2 *Inputs/Outputs*

2.3.2.1 Embedded Flip-Flops

In most of the designs, the inputs are synchronised to be handlled properly and avoid metastable states. The D flip-flops used to achieve this synchronisation often have an *enable* input. This *enable* input prevents new data to be sampled by the D flip-flop if it is driven low. By controlling this *enable* input, it is then possible to prevent the design from receiving new data from its inputs.

For example, for most of the FPGAs, the input/output blocks embed this type of D flip-flops. In order to specifically use this D flip-flop, some directives should be inserted in the design.

For Altera devices:

```
ATTRIBUTE useioff       : BOOLEAN;
ATTRIBUTE useioff OF e  : SIGNAL IS true;
```

For Xilinx devices:

```
ATTRIBUTE IOB        : STRING;
ATTRIBUTE IOB OF e  : SIGNAL IS''TRUE'';
```

For Lattice devices:

```
USE DIN TRUE CELL ''e'';
USE DOUT TRUE CELL ''e'';
```

The advantage of such technique is to reuse existing elements of the design. By using flip-flops which are already implemented, the overhead is very limited. It also has a strong impact on design operation, since it prevents new data to be loaded. However, it requires a specific type of flip-flop, since an *enable* input is necessary.

2.3.2.2 Fuses/Anti-fuses

In 2014, Basak et al. also proposed to act on the inputs of a circuit to get it to operate properly or not [9]. They propose to integrate anti-fuses in the chip's pins. Those anti-fuses are blown or not according to an authentication key. If the wrong key is supplied, then the wrong fuses are blown and the device is not usable. After the fuses are blown, the correct inputs and outputs are accessible and the device operates normally.

An interesting feature here is that if a system integrator obtains an integrated circuit on which fuses are already blown, then it is obviously a refurbished one. Therefore, such scheme also helps in fighting other types of threats on design intellectual property.

2.3.3 Processor

2.3.3.1 Processor's Programme Counter

Among complex systems, some integrate a soft-core processor in the FPGA fabric in order to execute programmes. Such processor is described in a hardware description language and instantiated. Altera Nios II [10] and Xilinx MicroBlaze [11] are examples of proprietary soft-core processors. An example of open-source soft-core processor is the Plasma CPU, available on the IP cores repository Opencores [12]. Moreover, some SoC actually include a wired processor. For example, recent Altera Cyclone V SoCs integrate a dual-core ARM Cortex-A9 processor.[1]

---

[1] https://www.altera.com/products/soc/portfolio/cyclone-v-soc/overview.html.

In order to disable a processor, acting on the programme counter, also called instruction pointer, is a very effective solution. The programme counter is a register that gives the address of the instruction being currently executed. Therefore, by controlling its value, it becomes possible to prevent new instructions from being executed. This can effectively halt the processor. Moreover, such halting can be set and released multiple times during the device's lifetime, allowing to achieve evaluation periods for instance. Therefore, acting on the programme counter is a versatile way to *license* the device. In case of counterfeiting or overbuilding, it can also obviously be used to render the processor unusable by permanently forcing the programme counter to a fixed value.

The detailed locking process is presented in Algorithm 1.

Algorithm 1: *Backup* and *locking* procedure

```
if locking request then
    if No branching or long instruction going on then
        if No branching or long instruction coming then
            PC_backup ← PC
Wait for ongoing instruction to finish then
PC ← "000...000"
instruction ← NOP
Locking completed
```

The first thing to do is to intercept the locking request. After that, it is important to verify that no problematic instruction is currently being executed. Indeed, if this is the case, then the return to normal operation is uncertain. Problematic instructions are long and branching instructions. Long instructions cannot be stopped during their execution and should terminate before. Similarly, during a branching instruction, the locking request should be postponed. This might cause the branching instruction to be skipped. The locking request should also be postponed if a problematic instructions is meant to be executed during the next clock cycle. In order to detect these instructions, the opcodes corresponding to problematic instructions can be read directly from the memory bus. These opcodes are provided by the processor's designer. This decision is based on practical experiments. It is important to ensure that a correct backup of the processor's current state is possible. Locking requests are not time-critical and can be postponed for several clock cycles to ensure processing integrity.

After that, the current value of the programme counter is stored in a dedicated register: PC_{backup}. At the end of the running instruction, which is not problematic, the programme counter is set to a nonfunctional value. In Algorithm 1, we took the example of the zero value ("000...000") but this can be different depending on the processor. The instruction register is set to *NOP*. This is to avoid executing the same instruction over and over when the processor is locked. It could modify its internal state and make the return to normal operation impossible.

Finally, the locking process is considered as completed.

In order to return to normal operation, the previous programme counter value should be restored. This is shown in Algorithm 2.

Algorithm 2: *Restore* procedure

if *unlocking request* **then**
  $PC \leftarrow PC_{backup}$
Wait for instruction to be loaded **then**
Unlocking completed

An interesting feature of this locking scheme is that it is fully reversible. Indeed, if the locking procedure has been followed properly, the instruction during which the locking occurred is not problematic. Therefore, the system can then be unlocked and start again without problem.

2.3.4 *Buses*

Buses integrity is crucial for correct communication between the different components of a system. Integrity can be more precisely defined in two terms: value and position. Thus data from a bus is sound if it has a correct value and it is correctly ordered.

Therefore, by acting on either the value or the position of the bus data, we can alter the bus operation. The first option is then to scramble the bus lines. The second option is to randomly mask the bus data.

For the subsequent Sects. 2.3.4.1 and 2.3.4.2, we assume that the bus is error free. The input value is identical to the output value during normal operation.

2.3.4.1 Deterministic Scrambling

A bus can be defined as the following function f:

$$f : \{0,1\}^n \rightarrow \{0,1\}^n \tag{2.3}$$

$$\forall x \in \{0,1\}^n : f(x) = x \tag{2.4}$$

It can thus be referred to as the identity function.

We define a deterministic scrambling function σ as :

$$\sigma : \{0,1\}^n \rightarrow \{0,1\}^m \; with \; n \geq m \tag{2.5}$$

$$\forall x \in \{0,1\}^n : \sigma(x) \neq x \tag{2.6}$$

However, such function can be heavy to implement. The requirement given in Eq. (2.6) is hard to fulfil for all x.

Therefore, we can define a relaxed version of the deterministic scrambling function σ_R as :

$$\sigma_R : \{0,1\}^n \rightarrow \{0,1\}^m \; with \; n \geq m \tag{2.7}$$

such that for *most of* the input values:

$$\sigma_R(x) \neq x \tag{2.8}$$

In fact, the relaxed version is sufficient for the usage we consider here. Indeed, disturbing a bus for even half of the input values is enough to render the overall system unusable.

The other point is to make the scrambling controllable by an additional input such that the scrambler can be turned on and off. A simple 2-to-1 n-bit multiplexer can be used to this end, selecting between the original bus data and the scrambled one. This is shown on Fig. 2.5.

From a practical point of view, implementing a scrambler is trivial. An n-bit circular shifter defined as $\sigma_R(x_i) = x_{i-1 \; mod \; n}$ and shown in Fig. 2.6 is efficient. It is only a relaxed deterministic scrambler since it does not alter the data if it consists or only 0 s or only 1 s.

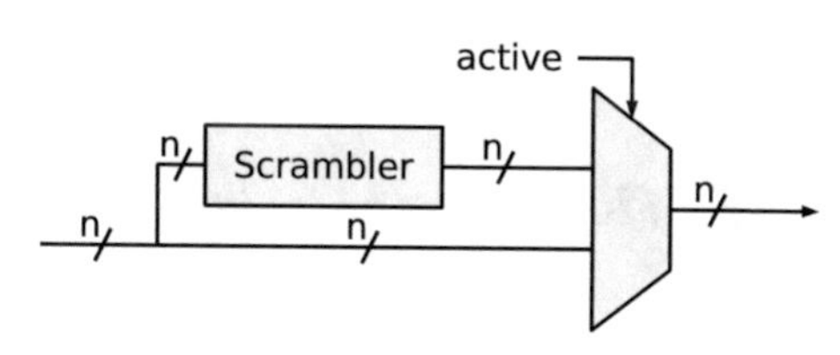

Fig. 2.5 Integration of the scrambler on an n-bit bus

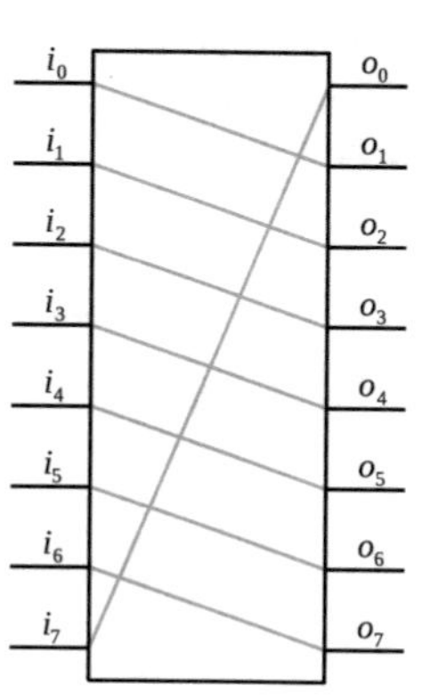

Fig. 2.6 8-bit circular shifter

However, there are even more trivial structures which can be used to scramble the bus. We do not give more details here as the chosen method is strongly dependent on the bus purpose.

Implementation details can require to add extra specifications to the scheme shown in Fig. 2.5. As shown on Fig. 2.1 of this chapter, a scrambler can be added to the address bus of the shared memory. This is a suitable choice, since reading from a wrong memory address disturbs the system heavily. However, writing to an unauthorised memory address could potentially alter the ability of the system to recover once the scrambler will be deactivated. For instance, the programme memory could be irremediably altered. Therefore, in this case, the bus should be scrambled only during read operations, not write.

2.3.4.2 Pseudo-random Masking

Another way to corrupt a bus is to act on the actual data which is transmitted through it. To this end, pseudo-random masking can be used.

In order to get pseudo-randomness, we use a linear feedback shift register (LFSR). Then, the shift register state bits are XOR-ed bitwise with the bus lines. For an n-bit bus, an n-bit shift register is used. If the feedback polynomial is carefully chosen, i.e. is primitive, a $2^n - 1$ clock cycles period can be obtained. This is shown in Fig. 2.7.

Similarly, the note made in the previous section about implementation-specific issues also applies to the pseudo-random masking scheme.

In order to reduce the power overhead induced by the LFSR, it can be clocked at a lower frequency then the nominal one of the design. Indeed, power consumption is proportional to the operating frequency.

2.3.5 *Finite State Machine*

The first way to modify the FSM is to add extra states before the original reset state. This is described in Sect. 2.3.5.1. The second option is to duplicate intermediate states to stop normal operation if the correct key is not provided. This is detailed in Sect. 2.3.5.2.

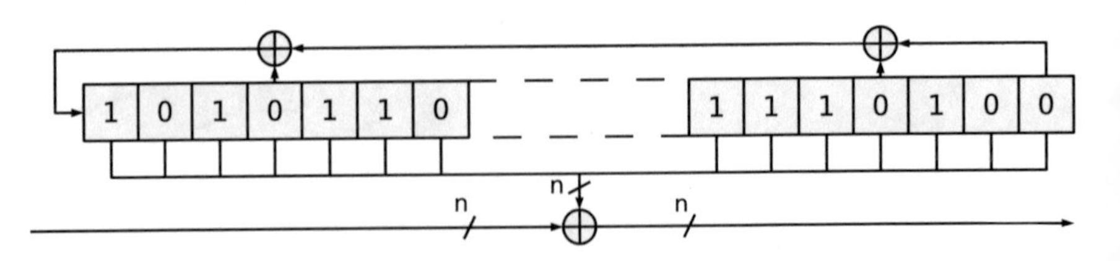

Fig. 2.7 Pseudo-random masking of a bus using an LFSR

2.3.5.1 Pre-reset States

The first possibility to modify the FSM is to add extra states before the original reset state [4, 13, 14]. This way, when the system is powered up or reset, it starts again in these extra states. In order to reach normal operation, the design must transact from one extra state to the other until it reaches the original reset state. If the state transitions of the extra states are only known to the original designer, then an attacker will not be able to reach the original reset state. The only possibility would be to explore all the extra states until the original reset state is reached.

The extra states can come at no cost if the original state machine is encoded in a way that so-called *don't care* states exist. If the FSM's states follow binary encoding, then an M-state FSM must use at least $\lceil log_2(M) \rceil$ D flip-flops to store the current state's value. If the number of flip-flops used is n, then there are $2^n - M$ states which are not used. These are *don't care* states. They can be used to encode the extra states.

A graphical representation of the modified FSM is shown in Fig. 2.8.

In this example, the original FSM includes five states. $\lceil log_2(5) \rceil = 3$, so three flip-flops are needed for state encoding. However, three flip-flops can encode $2^3 = 8$ states. Therefore, the three *don't care* states can be used as pre-reset states.

The new reset state is $S'0$. In order to transact to the original reset state $S0$, the correct values for K_0, K_1 and K_2 should be sent. If one key bit is wrong, then $S'0$ is reached again.

The advantage of such technique is to have low overhead since it makes use of *don't care* states. It has several drawbacks though. First of all, from a security point of view, this locking scheme exhibits a key even though it is not secure on its own. This can be misleading and get the designer to consider the scheme secure. However, security should rely on a cryptographic primitive. Second of all, transitions from one state to the other could be detected by the transient power consumption of switching flip-flops which encode the current state. Thus, finding the right key becomes trivial.

One option explored to make the scheme more secure is to initialise the state flip-flops to a random value, given by the response of a PUF to a specific challenge [4].

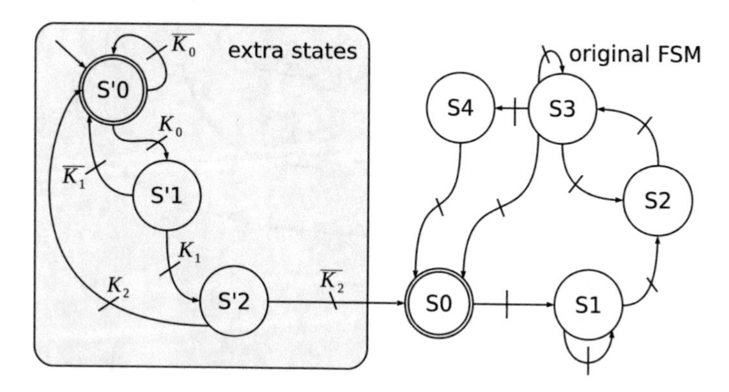

Fig. 2.8 Pre-reset states with key $(K_0K_1K_2) = 110$

Since only the designer knows the PUF's challenges/responses pairs, only he can find out the start-up state associated with a challenge. Therefore, in order to reach the original FSM, he must provide the system integrator with the right sequence of inputs to provide to the FSM. However, this only makes each FSM instance behave differently. It does not account for the two drawbacks previously described. Furthermore, it does not consider the variability of PUF's responses. If the PUF's response differs, the start-up state expected by the designer is different than the actual one of the powered-up device. Thus, the designer cannot provide the appropriate sequence of inputs to unlock the circuit.

Another option to modify the FSM is to duplicate specific states. This is described in the following section.

2.3.5.2 Duplicated States

Similarly, it is possible to use *don't care* states to duplicate some intermediate states. This is described in [15] and shown in Fig. 2.9. In this example, state *S*2 is duplicated.

The transitions from *S*1 to one of the duplicated states *S*21, *S*22, *S*23 and *S*24 is controlled by the output of a PUF, called random unique block in [15]. After that, in order to transition to the next state, here *S*3, a specific key must be applied to the FSM's inputs. This key is associated with the PUF's response and known only by the designer. If the wrong key is applied, the FSM does not transition to the next state and remains locked.

The advantages and drawbacks of this method are the same as the ones describes in Sect. 2.3.5.1.

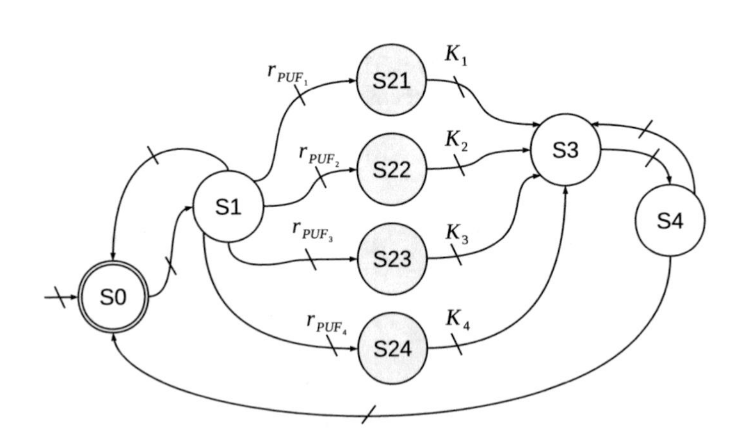

Fig. 2.9 Duplicated states S21, S22, S23 and S24

2.4 Implementation on FPGA and Results

We implemented the on-chip locks proposed in Sect. 2.3 on FPGA. We first give the cost for all solutions in terms of hardware resources. We then them implement on reference designs to estimate the implementation overhead. All the results are given with optimisation for lowest area. We used Quartus II 13.1 and ISE 13.4 for synthesis.

2.4.1 Hardware Resources

The experimental results obtained are given in Table 2.2. The implementation was carried out on two FPGA families: Altera Cyclone III and Xilinx Spartan 3. They are provided as the number of 4-input look-up tables (LUTs) and D flip-flops used for the implementation.

First, we can see that logic resources usage is very low for all the locks, except for the PLL reconfiguration. Indeed, in this case, the heaviest module is the one responsible for achieving the reconfiguration. It is provided by the FPGA manufacturer and can hardly be modified or optimised.

Conversely, all the other locks require very few logic resources. The most lightweight one consists in acting on the *enable* input of the input/output flip-flops. For buses, we give the required resources in terms of the bus width. They are always proportional to the bus width. When extra states are added to the FSM, either as pre-reset or duplicated states, the resources overhead grows logarithmically with respect

Table 2.2 Implementation results of on-chip locks alone on Altera Cyclone III and Xilinx Spartan 3

Modified feature	On-chip lock	#4-input LUTs	#D flip-flops
Clock circuitry	Reconfigurable PLL[a] (+ control FSM)	247(+55)	118(+18)
	Clock-gating module	9	6
Inputs/outputs	Inputs/outputs DFF enable	0	0
Interconnection bus	Deterministic scrambling[b]	8n/5	0
	Pseudo-random masking[b]	n	n
FSM	Pre-reset states[c]	log(n)	log(n)
	Duplicated states[c]	log(n)	log(n)

[a]Only on Cyclone III, not available on Spartan 3
[b]Of an *n*-bit bus
[c]For *n* extra states

Table 2.3 Required resources to implement the original designs

Unmodified	Altera Cyclone III		Xilinx Spartan 3	
design	#4-input LUTs	#D flip-flops	#4-input LUTs	#D flip-flops
Ethernet controller	275	108	357	99
Plasma CPU	2395	452	2901	394

to their number. We did not take into account here the possibility to reuse *don't care* states. This would reduce the required resources even further.

Modifying the programme counter is a specific process for each processor. This is detailed in Sect. 2.4.2.2 for the Plasma CPU.

2.4.2 *Reference Designs*

We then propose to implement the locks on two reference designs: an Ethernet controller and a soft-core processor. Both designs are available on the Opencores website [12]. For comparison, we give the resources required to implement the original designs in Table 2.3.

2.4.2.1 Ethernet Controller

The first reference design is an Ethernet controller. It is a fairly small design, mainly consisting in an FSM.

We first modified the clock circuitry. On the one hand, the clock-gating module requires 3 % more combinational fabric and 6 % more D flip-flops. The resources overhead is then rather low. On the other hand, implementing a reconfigurable PLL doubles the required resources, and is clearly not practical.

As expected, acting on the D flip-flop *enable* input requires no extra resources.

When acting on the bus, scrambling is cheaper than masking. We implemented them both on a 32-bit bus. Scrambling requires extra combinational logic, around 9 % more. On the other hand, pseudo-random masking needs D flip-flops to implement the LFSR. Therefore, the associated overhead is quite high, around 30 % more D flip-flops.

Finally, we also modified the FSM in both ways. First, we added 32 pre-reset states. Even if it only requires one extra flip-flop, the combinational logic handling the transitions between the extra states is heavy, and requires almost 25 % more resources. Then, we duplicated one of the state 32 times. Similarly, only one extra flip-flop was added to the design but the requirement for combinational logic exceeds 70 % here, which is excessive.

Table 2.4 Implementation of on-chip locks on the Ethernet controller

Modified	Altera Cyclone III		Xilinx Spartan 3	
design	#4-input LUTs	#D flip-flops	#4-input LUTs	#D flip-flops
Clock-gating module	284 (+3 %)	114 (+6 %)	367 (+3 %)	105 (+6 %)
PLL reconfiguration	522 (+90 %)	226 (+109 %)	*Not available*	
Inputs/outputs DFF enable	275 (+0 %)	108 (+0 %)	357 (+0 %)	99 (+0 %)
Deterministic scrambling[a]	297 (+8 %)	108 (+0 %)	388 (+9 %)	99 (+0 %)
Pseudo-random masking[a]	313 (+14 %)	140 (+30 %)	391 (+9 %)	131 (+32 %)
32 pre-reset states	343 (+25 %)	109 (+1 %)	425 (+19 %)	100 (+1 %)
Duplicated state (x32)	493 (+80 %)	109 (+1 %)	612 (+71 %)	100 (+1 %)

[a]For a 32-bit bus

All the results of the implementation on the Ethernet controller are shown in Table 2.4.

For a design of this size, we can then estimate that only input/output locking is suitable. Adding a clock-gating module can be also considered, since the overhead is still rather low. All the other modifications lead to an important overhead.

2.4.2.2 Plasma CPU

A larger design is now presented, the soft-core processor Plasma CPU.

Here, the clock-gating module is even cheaper in terms of resources. However, PLL reconfiguration remains expensive, with 13 % more combinational resources and 30 % more D flip-flops.

Like before, using integrated input/output D flip-flops adds no logic resources.

Modifying the bus becomes affordable with this kind of large designs. Scrambling it in a deterministic way induces almost no overhead. Pseudo-random masking leads to low overhead, below 8 % Therefore, it can be implemented as a powerful way to disturb the bus data.

Since the Plasma CPU does not comprise an FSM, pre-reset and duplicated states could not be implemented on this design.

Finally, being able to control the programme counter value is also quite expensive and requires 10 % extra resources. However, this was implemented on an already existing design. We assume it could be implemented in a more lightweight way if this feature was taken into account during the design phase.

All the implementation results are provided in Table 2.5.

Table 2.5 Implementation of on-chip locks on the Plasma CPU

Modified design	Altera Cyclone III		Xilinx Spartan 3	
	#4-input LUTs	#D flip-flops	#4-input LUTs	#D flip-flops
Clock-gating module	2399 (+0.17 %)	458 (+1 %)	2932 (+1 %)	400 (+2 %)
PLL reconfiguration	2697 (+13 %)	588 (+30 %)	*Not available*	
Inputs/outputs DFF enable	2395 (+0 %)	452 (+0 %)	2901 (+0 %)	394 (+0 %)
Deterministic scrambling[a]	2430 (+1 %)	452 (+0 %)	2894 (+0 %)	394 (+0 %)
Pseudo-random masking[a]	2446 (+2 %)	484 (+7 %)	2927 (+1 %)	426 (+8 %)
Programme counter halt	*Not implemented*		3186 (+10 %)	428 (+9 %)

[a]For a 32-bit bus

On larger designs, implementing more complex locks is possible. The associated overhead is limited, and even multiple locks could be integrated.

2.5 Discussion: Partial Locking

For some of the features presented previously, it is possible to achieve partial locking. Partial locking can refer to the following features:

- Lower performance: lower operating frequency...
- Less features: reduced instruction set...
- Limlited period of use: trial period.

However, the system is still perfectly functional.

Such partial locking feature is very interesting from the point of view of the designer. In fact, it allows the designer to provide the design in evaluation mode. This way, the design can be thoroughly tested by the future user before buying it. With the increasing number of designs provided as IP cores, this allows to support a business model similar to the one used in for software distribution. The product is first provided as an evaluation version and can then be fully unlocked to perform optimally. This offers interesting flexibility to the licensing model.

What is important to partially lock a design is to choose a locking scheme that provides some granularity. Among the modifications we have shown in the previous sections, acting on the clock circuitry is the only one able to achieve this. Indeed, as said before, the clock frequency is directly related to the performance of the design. Therefore, acting on the clock frequency using a clock-gating module or by dynamically reconfiguring the PLL is a way to tune system performance.

In order to implement this kind of functionality, a controller should also be implemented to handle the different states in which the system can operate: locked, evaluation or unlocked. Different commands and the associated unique keys are controlling the transitions from one state to another. It leads to additional resources overhead, which must be taken into account.

2.6 Conclusion

This chapter shows how some features, already existing in most electronic designs, can be turned into powerful on-chip locks. Since they are already present, using them lead to low resources overhead. We provide different techniques about how to effectively implement these locks. We also give details on partial locking. This can help to achieve a more flexible licensing model for IP cores, allowing to provide devices in evaluation mode.

The most suited way to modify a design to make it lockable seems to be the modification of the clock circuitry. It is lightweight and can effectively tune the performance of the design.

References

1. Frontier-Economics, Estimating the global economic and social impacts of counterfeiting and piracy, in *Business Action to Stop Counterfeiting and Piracy (BASCAP)*, Tech. Rep. (2011)
2. C. Gorman, Counterfeit chips on the rise. IEEE Spectrum **49**(6), 16–17 (2012)
3. U. Guin, K. Huang, D. DiMase, J.M. Carulli, M. Tehranipoor, Y. Makris, Counterfeit integrated circuits: a rising threat in the global semiconductor supply chain. Proc. IEEE **102**(8), 1207–1228 (2014)
4. Y. Alkabani, F. Koushanfar, Active hardware metering for intellectual property protection and security, in *USENIX Security* (Boston MA, USA, 2007), pp. 291–306
5. R. Karri, J. Rajendran, K. Rosenfeld, M. Tehranipoor, Trustworthy hardware: identifying and classifying hardware trojans. Computer **43**(10), 39–46 (2010)
6. L. Bossuet, D. Hély, SALWARE salutary: hardware to design trusted IC, in *Workshop on Trustworthy Manufacturing and Utilization of Secure Devices, TRUDEVICE* (2013)
7. B. Colombier, L. Bossuet, Functional locking modules for design protection of intellectual property cores, *IEEE International Symposium on Field-Programmable Custom Computing Machines* (Vancouver, Canada, 2015), p. 233
8. Altera. Implementing fractional pll reconfiguration with altera pll and altera pll reconfig ip cores (2015)
9. A. Basak, Y. Zheng, S. Bhunia, Active defense against counterfeiting attacks through robust antifuse-based on-chip locks, in IEEE 32nd *VLSI Test Symposium* (Napa CA, USA, 2014), pp. 1–6
10. Altera. Nios, https://www.altera.com/products/processors/overview.html
11. Xilinx. Microblaze, http://www.xilinx.com/products/intellectual-property/microblazecore.html
12. Opencores, http://www.opencores.org

13. R.S. Chakraborty, S. Bhunia, Security against hardware trojan through a novel application of design obfuscation, in *International Conference on Computer-Aided Design* (ACM, 2009), pp. 113–116
14. R.S. Chakraborty, S. Bhunia, HARPOON: an obfuscation-based soc design methodology for hardware protection. IEEE Trans. Comput. Aided Des. Integr. Circuits Syst. **28**(10), 493–1502 (2009)
15. Y. Alkabani, F. Koushanfar, M. Potkonjak, Remote activation of ICs for piracy prevention and digital right management. in *IEEE/ACMInternational Conference on Computer-aided Design*, (Beijing, China, 2007), pp. 674–677

Chapter 3
Logic Modification-Based IP Protection Methods: An Overview and a Proposal

Brice Colombier, Lilian Bossuet and David Hély

3.1 Introduction and Context

Design data protection schemes can be classified into two categories. *Passive* ones detect that counterfeiting of over-usage took place, but do not stop it. Conversely, *active* protection schemes actually prevent the infringement to occur in the first place. They do so by modifying the design in order to make it resilient to such threats. Chapter 2 gives an overview of the available features in most electronics design that can be turned into powerful locks. The current chapter focuses on protection means that require a modification of the combinational logic.

The first feature which can be achieved by logic modification is to controllably disturb the outputs. This allows the designer to make the circuit unusable. In order to return to normal operation, a "key" must be provided to the circuit. We use the word "key" here as a generic term, not as a cryptographically strong sequence of digits. This key is provided by the designer, who can therefore record how many times the key has been delivered, and how many instances of the circuit are activated. Such "count of the produced ICs" is called hardware metering [1]. It can be achieved by different means, which are presented in the next section.

The second feature which can justify logic modification is to slow down reverse-engineering. By adding extra logic gates, recovering the circuit functionality from a high-resolution picture [2] of the layout or a netlist can become extremely difficult. Logic obfuscation is one of the ways to do this, and is also presented.

B. Colombier (✉) · L. Bossuet
Hubert Curien Laboratory, UMR CNRS 5516, University of Lyon, Saint-Étienne, France
e-mail: b.colombier@univ-st-etienne.fr

L. Bossuet
e-mail: lilian.bossuet@univ-st-etienne.fr

D. Hély
LCIS, Grenoble Institute of Technology, Valence, France
e-mail: david.hely@lcis.grenoble-inp.fr

L. Bossuet and L. Torres (eds.), *Foundations of Hardware IP Protection*,
DOI 10.1007/978-3-319-50380-6_3

One of the key features of all logic modification-based protection schemes is the selection of the sites to modify. Those sites are the ones on which extra gates will be inserted. Different techniques can be used to this end, such as random selection [3], fault-analysis [4], etc. A trade-off between efficiency and computation time must be done by the designer. For example, finding the best place to insert a masking gate can be very time consuming, as shown in [4]. A novel technique which uses graph-analysis methods is presented. It selects the sites to modify orders of magnitude faster than fault analysis-based techniques, yet achieving better outputs disturbance than simple random selection.

Finally, all the schemes mentioned above need to be integrated in a complete design protection module. Indeed, even though many previous works try to exhibit security features in their protection schemes, such security can only be reached by using a dedicated cryptographic function. This is discussed in more details in the final section.

This chapter is organized as follows. In Sect. 3.2, we provide a formal framework for logic modification-based protection schemes by defining logic encryption, logic obfuscation, logic masking, and logic locking and give examples for each. In Sect. 3.3, we present a new graph-based algorithm that selects the optimal nodes to be modified to achieve logic locking of a combinational netlist. In Sect. 3.4, we present the results of implementation, specifically the logic resources overhead and analysis time. In Sect. 3.5, we evaluate the proposed method and develop associated metrics. In Sect. 3.6 we describe a threat model and perform a security analysis of the protection schemes considered. In Sect. 3.7, we discuss design considerations. In particular, we emphasize the need to introduce a cryptographic primitive to ensure security, and to not rely on the logic/masking module to fulfill this objective.

3.2 A Formal Foundation for Logic Protection Schemes

An increasing number of works are trying to find a way to protect the intellectual property of IP designers and fabless IC designers by acting on combinational logic. Unfortunately, most of these works make incorrect use of the terminology, i.e., *logic encryption*, *logic obfuscation*, *logic masking* and *logic locking* are used without a formal definition. This chapter takes the opportunity to propose a formal foundation for logic protection schemes. In this section, we provide formal descriptions and definitions of the logic protection schemes in order to strictly evaluate their different contributions to the literature. In all the following subsections, the original (not protected) n-input, l-output logic function is formalized by a Boolean function $f\{0,1\}^n \rightarrow \{0,1\}^l$.

3.2.1 Logic Encryption

The term "logic encryption" is used when a specific symmetric encryption function ξ_f over $GF(2^l)$ is applied to f. Formally, it is not *logic encryption*. The term is not specific. *Encryption of the Boolean function* f is the correct expression. The result of this encryption is the Boolean function $f'\{0,1\}^n \rightarrow \{0,1\}^l$. f' is given by the following expression, where k is the secret key:

$$f' = \xi_f(f, k)$$

ξ_f is a symmetric encryption function if and only if an inverse function ψ_f exists that uses the same secret key k for decryption, and is defined as follows:

$$\psi_f(f') = \psi_f(\xi_f(f, k), k) = f \tag{3.1}$$

Functions ξ_f and ψ_f must meet the following requirements:

$$\forall (k_i, k_j) \in (\{0,1\}^m, \{0,1\}^m), k_i \neq k_j$$

$$\xi_f(f, k_i) \neq \xi_f(f, k_j) \tag{3.2}$$

$$\psi_f(\xi_f(f, k_i), k_i) \neq \psi_f(\xi_f(f, k_i), k_j) \tag{3.3}$$

Functions ξ_f and ψ_f also have to satisfy the following requirements, where *Corr* is the function that computes Pearson's correlation coefficient.

$$\forall k \in \{0,1\}^m : Corr(\xi_f(f, k), f) \simeq 0 \tag{3.4}$$

$$\forall k \in \{0,1\}^m : Corr(\psi_f(\xi_f(f, k), k), \xi_f(f, k)) \simeq 0 \tag{3.5}$$

One of the consequences of the last expression is that the mean of the Hamming distance between the input and the output of the encryption/decryption functions is close to 50 % (ideally exactly 50 %) as described by the following expressions when the mean of the Hamming distance is computed for all the inputs of the Boolean function f:

$$\forall k \in \{0,1\}^m : \frac{\sum HD(\xi_f(f\{0,1\}^n, k), f\{0,1\}^n)}{2^n - 1} \simeq 50\,\% \tag{3.6}$$

$$\forall k \in \{0,1\}^m : \frac{\sum HD(\psi_f(\xi_f(f\{0,1\}^n, k)), \xi_f(f\{0,1\}^n))}{2^n - 1} \simeq 50\,\% \tag{3.7}$$

Some works [4–6] consider this last property as proof of security. This is a mistake, since it is possible to obtain the same result with a function that does not achieve

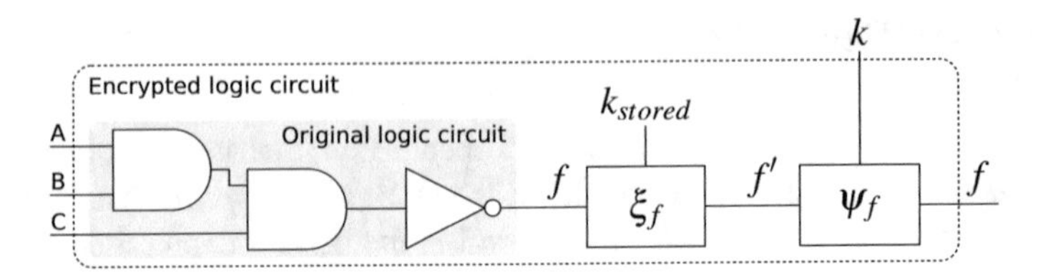

Fig. 3.1 Example of logic encryption

encryption. For instance, inverting the first $n/2$ bits of the output of f leads to a 50 % Hamming distance. Similarly, inverting every input of odd order leads to the same result. In both cases, the mean of the Hamming distance as described in (3.6) is equal to 50 % but the correlation defined in (3.4) is not zero.

These works are presented as "logic encryption," even though this is absolutely not the case. The authors of these works defined "logic encryption" as: "*logic encryption hides the functionality and the implementation of a design by inserting some additional gates called key-gates into the original design*" [5]. With this definition, logic encryption does not respect the expressions (3.1) to (3.7). Consequently, we claim that all works presented as "logic encryption" are inaccurate because in fact, they only propose to *mask* the logic functionality. The security level of such masking functions is very low compared with proper encryption.

A didactic example of true "logic encryption" is given by considering the following 3-input Boolean function $f\{0,1\}^3 \rightarrow \{0,1\}^1$:

$$f(A,B,C) = \overline{A.B.C}$$

Figure 3.1 is a diagram of the encrypted logic circuit. This includes the original logic circuit which computes the Boolean function f, the encryption function ξ_f which computes the encrypted Boolean function f' using an embedded secret key k and the decryption function ψ_f which outputs the correct result of the Boolean function f if and only if the correct key k is applied on the external key input.

This didactic example shows that the area overhead of true logic encryption is always prohibitive since it requires the implementation of encryption and decryption functions. Note that the security level of such a protection depends on the key size. Nowadays a secure implementation of a symmetric cipher has to use at least a 128-bit key. All protection schemes that include a secret key that has only a few bits fail to provide the designer with any security because of the feasibility of a brute force attack.

3.2.2 *Logic Obfuscation*

Logic obfuscation comes from the field of computer science in which developers wish to protect source codes against unauthorized reading and understanding.

The following definition of code obfuscation is proposed by Hachez [7]: *Transform a program P into another program P' harder to reverse engineer with the same observable behavior. If P fails to terminate or terminates with an error, then P' fails to terminate or terminates with an error. Otherwise, P' must terminate and produce the same output as P.* Hardware obfuscation consists in applying this definition to the hardware field, by changing the logic, FSM, or other part of a design without changing the system behavior.

When the logic part of a circuit is obfuscated, a design modification γ_f is applied to f. The result of this design modification is the Boolean function $f''\{0,1\}^n \rightarrow \{0,1\}^l$.

$$\gamma_f(f) = f''$$

The function γ_f must meet the following requirement for any input $x \in \{0,1\}^l$:

$$\forall x \in \{0,1\} : f''(x) = f(x) \tag{3.8}$$

Some works present logic obfuscation but do not fulfill requirement (3.8) [8, 9]. Most of these works use a secret key that changes the behavior of the original logic function. These works are typical cases of logic masking, which is presented in Sect. 3.2.3.

It is possible to try to perform obfuscation at the logic-gate level but this usually implies a large overhead. Indeed, obfuscation techniques aim to increase reverse-engineering time. The time is at least linear with the area [10]. Increasing the area increases the time needed for reverse engineering. As a consequence, the main design modification rule for obfuscation is to not follow the usual design rules for efficient implementation of a Boolean function. Usually, laws and theorems of Boolean logic are applied to Boolean functions in order to reduce the number of gates (i.e., the area) of the final hardware implementation. To obfuscate an implementation of a Boolean function, these laws and theorems are followed in the opposite way, i.e., they increase the size of the hardware implementation.

Two strategies are used in the first step of obfuscation: *develop* and *obscure*. To develop a Boolean function, the designer can use the canonical disjunctive normal form (also called *min-term* canonical form) in which the Boolean function is represented and implemented as a sum of *min-terms*. As a didactic example, let us consider the following 3-input Boolean function $f\{0,1\}^3 \rightarrow \{0,1\}^1$:

$$f(A,B,C) = \overline{A.B.C}$$

This Boolean function could be developed using the following canonical disjunctive normal form (first obfuscation step).

$$f''(A,B,C) = A.\overline{B}.\overline{C} + \overline{A}.\overline{B}.\overline{C} + \overline{A}.B.\overline{C} + A.B.C + A.\overline{B}.C + \overline{A}.\overline{B}.C + \overline{A}.B.C$$

f and f'' follow requirement (3.8). Figure 3.2a, b show the logic diagrams of the two functions with only 2-input AND and OR gates and inverters (other types of gates could also be used).

In order to obscure a Boolean function, the designer can apply to f'' some of the Boolean logic laws (absorption, complementary, common identities, etc.) and DeMorgan's theorem to increase the number of gates used in the hardware implementation. For example, by also using some redundant logic operations, f'' is described by the following Boolean expression:

$$\begin{aligned} f''(A,B,C) =& A.\overline{B} + \overline{A}.\overline{B} + \overline{A}.B + \overline{B}.\overline{C} + \overline{A}.\overline{C} + A.C + \overline{B}.C + \overline{A}.C + B.C + \overline{A} \\ &+ \overline{B} + C + \overline{A.B} + A \oplus C + A \oplus B + \overline{A.\overline{C}} + \overline{B.\overline{C}} \end{aligned}$$

Again f and f'' follow requirement (3.8). Figure 3.2c shows the logic diagram of f'' after this second step of obfuscation. The designer can also insert dummy logic to further increase the reverse engineering effort.

Table 3.1 shows the logic resources required for each logic circuit in Fig. 3.2. For each circuit, the number of gates is shown for each type (inverter, 2-input *and* gate, 2-input *or* gate and 2-input *xor* gate), along with the gate equivalent metric. The area overhead is given for the two hardware implementations of f''. As mentioned above, the increase in reverse-engineering time for each obfuscated logic circuit (in comparison with the original logic circuit) is supposed to be equal to the area overhead. For example, the time required to reverse engineer circuit shown in Fig. 3.2c is 14.58 times greater than the time required to reverse engineer the original circuit.

Due to the high area overhead, such logic obfuscation is not suitable for most applications. Moreover, the hardware design of the obfuscated circuit has to be performed by hand to avoid logic optimization by the synthesis tool. It is possible to mix a light logic obfuscation with obfuscation at another level. Indeed, hardware obfuscation is also possible at the HDL [11, 12] or layout levels [13, 14].

The above description of logic encryption and logic obfuscation allows us to affirm that none of the published works that present "logic encryption" or "logic obfuscation" meet the formal requirements of these two techniques. Most of these works in fact describe "logic masking" or "logic locking." In the remainder of this section we present logic masking and logic locking techniques.

3.2.3 *Logic Masking*

Logic masking consists in inserting *xor* or *xnor* gates in the data path of the logic circuit of a Boolean function in order to change the logic behavior of the circuit if the wrong masking key is applied. It was first proposed in [3]. Let us consider that a Boolean function $f\{0,1\}^n \rightarrow \{0,1\}^l$ could be represented as a set of i Boolean subfunctions $\{f_0, f_1, \dots, f_{i-1}\}$. Logic masking of the Boolean function f by using the i-bit

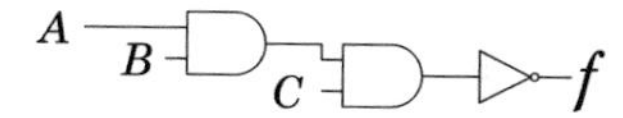

(a) Original Boolean function implementation

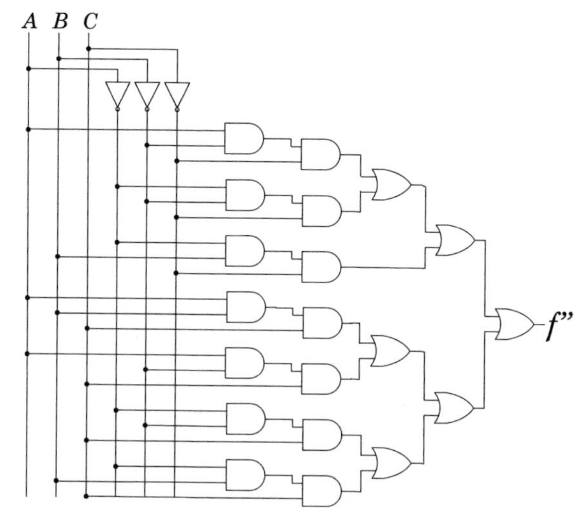

(b) Boolean function implementation after a first step of logic obfuscation

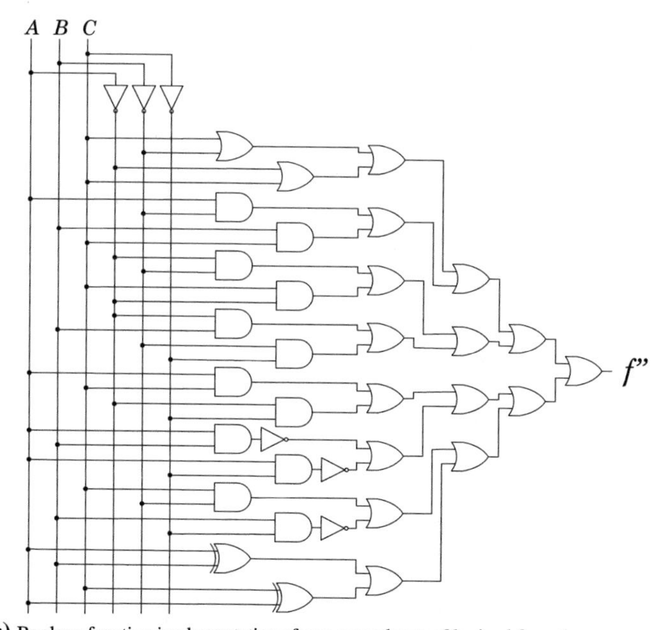

(c) Boolean function implementation after a second step of logic obfuscation

Fig. 3.2 **a** Original boolean function implementation, **b** Boolean function implementation after a first step of logic obfuscation, **c** Boolean function implementation after a second step of logic obfuscation

Table 3.1 Logic resources requirements and timing overhead for reverse-engineering of the circuits described in Fig. 3.2

Boolean function	Logic circuit	#Logic gates				Gate equivalent	Area/reverse-engineering time overhead
		inv	*and*	*or*	*xor*		
f	Figure 3.2a	1	2			4.01	–
f after first step of obfuscation	Figure 3.2b	3	14	6		35.41	+883 %
f after second step of obfuscation	Figure 3.2c	6	12	17	2	58.47	+1458 %

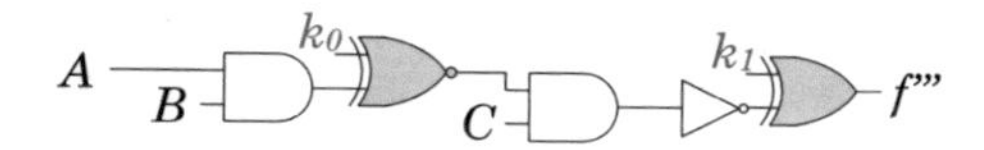

Fig. 3.3 Example of logic masking

masking key $k = \{k_0, k_1, \dots, k_{i-1}\}$ is described by the following expression, where f''' is a Boolean function $f\{0,1\}^n \rightarrow \{0,1\}^l$ and $\ominus$ is the *xor* or *xnor* Boolean operator:

$$f''' = \{f_0 \ominus_0 k_0, f_1 \ominus_1 k_1, \dots, f_{i-1} \ominus_{i-1} k_{i-1}\}$$

$$\forall j \in [0, i-1] \begin{cases} \text{if } \ominus_j \equiv xor \rightarrow k_j = 1 \rightarrow f_j \ominus k_j = f_j \\ \text{if } \ominus_j \equiv xnor \rightarrow k_j = 0 \rightarrow f_j \ominus k_j = f_j \end{cases} \tag{3.9}$$

The correct masking key k is found by using the laws in (3.9), and considering the type of inserted gate. As a didactic example, let us consider the following 3-input Boolean function $f\{0,1\}^3 \rightarrow \{0,1\}^1$:

$$f(A,B,C) = \overline{A.B.C}$$

This Boolean function could also be described by the following expression:

$$\begin{cases} f(A,B,C) = f_1(f_0(A,B), C) \\ f_0(X,Y) = X.Y \\ f_1(X,Y) = \overline{X.Y} \end{cases}$$

A didactic example of logic masking of the Boolean function f is given in Fig. 3.3, where $\ominus_0$ is an *xnor* gate and $\ominus_1$ is an *xor* gate. According to the laws in (3.9), we can determine the correct masking key $k = \{0, 1\}$ needed to obtain the original logic behaviour. In Fig. 3.3, additional masking gates are in gray.

Efficient insertion of the masking scheme has to be achieved without reducing performance (mainly by limiting the insertion of gates on the critical path) or increasing area overhead (by limiting the number of additional gates without using too few bits for the masking key k). For example, works presented in [4, 15] propose to use heuristics to reduce overhead.

3.2.4 Logic Locking

Logic locking allows the designer to insert *or*, *and*, *nor* or *nand* gates in the data path of the logic circuit of a Boolean function in order to lock the output to a fixed logic level (0 or 1) if the wrong unlocking key is applied. Let us consider that a Boolean function $f\{0,1\}^n \rightarrow \{0,1\}^l$ can be represented as a set of i Boolean subfunctions $\{f_0, f_1, \dots, f_{i-1}\}$. Logic locking of the Boolean function f by using the i-bit unlocking

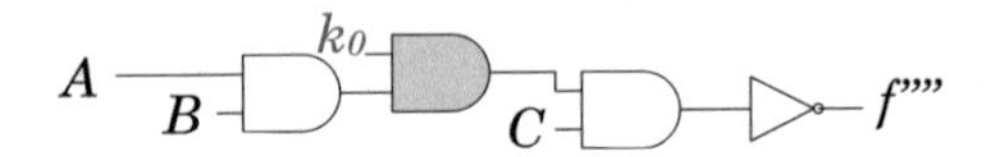

Fig. 3.4 Example of logic locking

word $k = \{k_0, k_1, \ldots, k_{i-1}\}$ is described by the following expression when f'''' is a Boolean function $f\{0,1\}^n \rightarrow \{0,1\}^l$ and $\odot$ is the *and* or *or* Boolean operator:

$$f''' = \{f_0 \odot_0 k_0, f_1 \odot_1 k_1, \ldots, f_{i-1} \odot_{i-1} k_{i-1}\}$$

$$\forall j \in [0, i-1] \begin{cases} \text{if } \odot_j \equiv and \rightarrow k_j = 1 \rightarrow f_j \odot k_j = f_j \\ \text{if } \odot_j \equiv or \rightarrow k_j = 0 \rightarrow f_j \odot k_j = f_j \end{cases} \tag{3.10}$$

The correct unlocking key k is found by using the laws in (3.10), and considering the type of inserted gate. As a didactic example, let us consider the following 3-input Boolean function $f\{0,1\}^3 \rightarrow \{0,1\}^1$:

$$f(A, B, C) = \overline{A.B.C}$$

This Boolean function could be expressed by the following expression:

$$\begin{cases} f(A, B, C) = f_1(f_0(A, B), C) \\ f_0(X, Y) = X.Y \\ f_1(X, Y) = \overline{X.Y} \end{cases}$$

A didactic example of logic locking of the Boolean function f is given in Fig. 3.4 where $\odot_0$ is an *and* gate. In this very simple example, only one gate is used to lock the logic behavior of the circuit. By following the laws in (3.10) we can determine the correct unlocking word $k = 1$ to obtain the correct behaviour. In Fig. 3.4 the additional locking gate is in gray.

Like for logic masking, the insertion of the locking gates has to be achieved without reducing performance and increasing area overhead. In the following section, we present a new method based on the graph analysis of an RTL netlist, which achieves efficient and secure logic locking.

Like in logic obfuscation and masking, it is possible to lock a circuit by acting on parts/levels other than the logic level. For example, recent works propose to lock the finite-state-machine [16, 17] or the input/output ports [18].

3.3 Proposed Graph Analysis-Based Logic Locking Scheme

As mentioned in Sect. 3.2.4, what we propose here is a new technique to select the nodes to include in the logic locking process. Indeed, since logic locking requires the insertion of extra logic gates, it is necessary to find the optimal place in the combinational netlist on which these extra gates should be inserted. According to the previously proposed definition, logic locking is the propagation of a fixed logic value from an internal node to one or several output(s). To achieve this, we need to identify sequences of gates that could propagate such a logic value. To this end, we represent the netlist as a graph. This representation is a convenient way of analyzing relations between logic gates and finding the optimal paths in a netlist that could propagate the logic locking value.

3.3.1 Implementation of Logic Locking

Before building the graph, we must identify the characteristics leading to the propagation of a locking value in a sequence of logic gates. First, it is worth noting that a specific *controlling value* exists for nonlinear logic gates. If this controlling value is applied to one of the logic gate's inputs, then the output is forced to a fixed, known value. For instance, setting one of the inputs of an *and* gate to 0 will set the output to 0. Table 3.2 summarizes the controlling values for the four 2-input nonlinear logic gates.

Next, for every node in the netlist, we define two values: V_{locks} and V_{forced}. V_{locks} is the controlling value of the gate that comes after this node. For instance, if a node is the input of an *or* gate, then $V_{locks} = 1$. V_{forced} is the value to which the node will be forced. For instance, if a node is the output of an *or* gate, $V_{forced} = 1$. It should be noted that in some cases $V_{locks} = \{0, 1\}$. This occurs if the node has a fan-out higher than one and spans gates with different controlling values. A node is useful for logic locking if it is forced to the controlling value of the following gate. Therefore, for sequences of nodes that can propagate a locking value, all the nodes meet the following criterion:

$$\textbf{Criterion 1}: V_{forced} \in V_{locks}$$

If Criterion 1 is verified for all the nodes in a sequence of nodes, then this sequence is able to propagate a locking value. In this case, forcing the first node to its controlling value will set all the nodes in the sequence at a fixed logic value. This is illustrated in Fig. 3.5.

Table 3.2 Controlling value and associated output value for 2-input nonlinear logic gates

Logic gate	Controlling value	Output value[a]
and	0	0
nand	0	1
or	1	1
nor	1	0

[a]when the controlling value is applied to one of the inputs

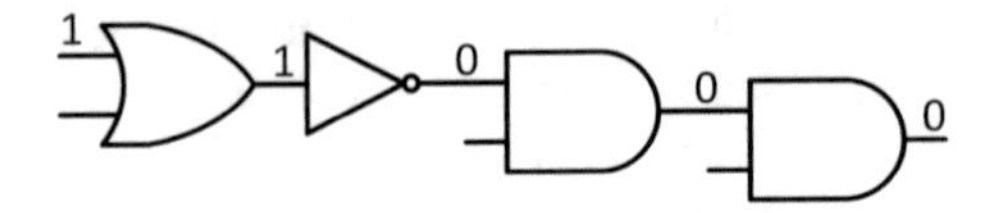

Fig. 3.5 Propagation of a locking value in a sequence of logic gates

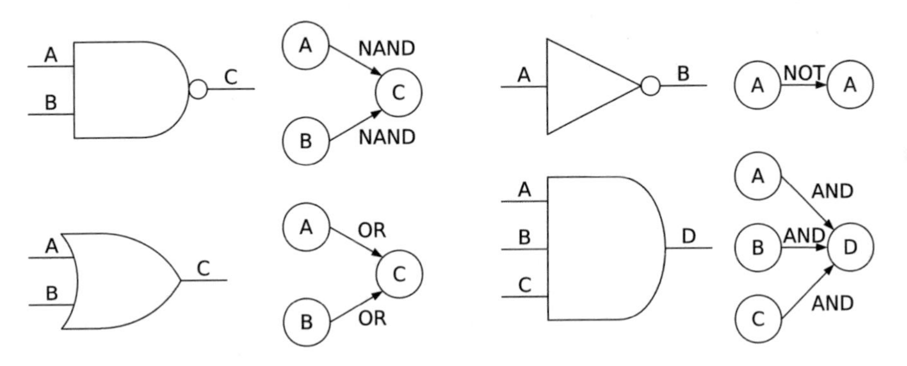

Fig. 3.6 Conversion from logic gates to graph elements

With this in mind, one can see how an output can be forced to a fixed logic value. By inserting logic gates at specific locations in the netlist, the designer will be able to set controlling values and force the outputs to a fixed value. The aim here is to select the most appropriate nodes, namely those at the beginning of sequences of gates like the one presented in Fig. 3.5. To achieve this aim, graph exploration techniques are used, and are presented in the following sections.

3.3.2 Graph Building

The original design file is an RTL description of the combinational netlist. The first step is to convert it into a directed acyclic graph. We chose to represent the netlist's nodes as vertices and the Boolean functions as edges. An example of conversion from logic gates to graph elements is shown in Fig. 3.6.

This is repeated for all logic gates of the netlist. A toy example of a netlist converted into a graph is shown in Fig. 3.7.

In order to identify which nodes satisfy criterion 1, V_{locks} and V_{forced} are computed for all the nodes in the netlist (i.e., all the vertices in the graph). This is done as follows: outgoing edges are used to compute V_{locks}, while incoming edges are used to compute V_{forced}. By convention, for the sake of the following computations, V_{locks} is set to $\{0, 1\}$ for the outputs. Table 3.3 shows V_{locks} and V_{forced} values computed for all the vertices of the graph shown in Fig. 3.7.

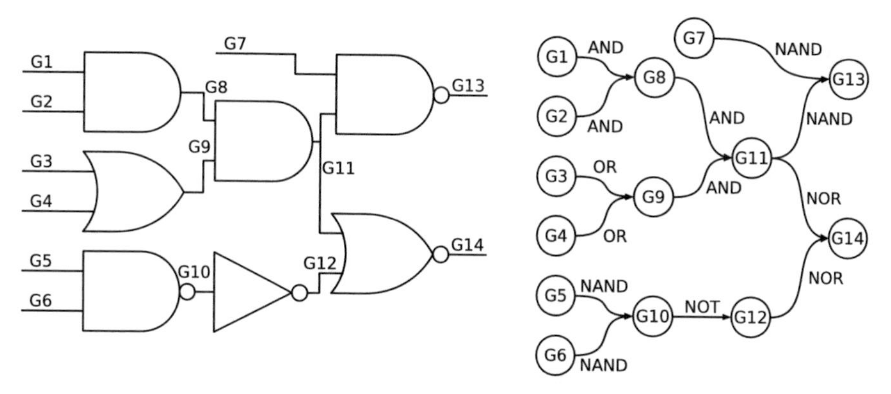

Fig. 3.7 Conversion from netlist to graph

Table 3.3 V_{locks} and V_{forced} values for all the nodes of the netlist shown in Fig. 3.7

Node	V_{forced}	V_{locks}	Node	V_{forced}	V_{locks}
G1	–	0	G8	0	0
G2	–	0	G9	1	0
G3	–	1	G10	1	0
G4	–	1	G11	0	$\{0, 1\}$
G5	–	0	G12	0	1
G6	–	0	G13	1	$\{0, 1\}$
G7	–	0	G14	0	$\{0, 1\}$

The next step is to identify which nodes cannot propagate the locking value. This means they do not fulfill criterion 1. If a node does not meet this criterion, its incoming edges are deleted. Thus in the previous example, incoming edges are deleted for G9, G10, and G12.

What is obtained at this stage is a highly disconnected graph, because the vast majority of vertices do not fulfill criterion 1. Since we want to achieve logic locking, connected components that do not contain any output must be removed from the graph. After applying this method to the graph in the previous example, we obtain the one shown in Fig. 3.8. The original netlist is shown too, and a path that can propagate a locking value is highlighted.

The final graph obtained at this stage comprises nodes that can all propagate a locking value to the output if they are forced to a specific logic value. Some of them, however, are better candidates, because they span a greater number of outputs or are more deeply integrated in the netlist. The selection algorithm used to identify the best nodes to act on is described in the following section.

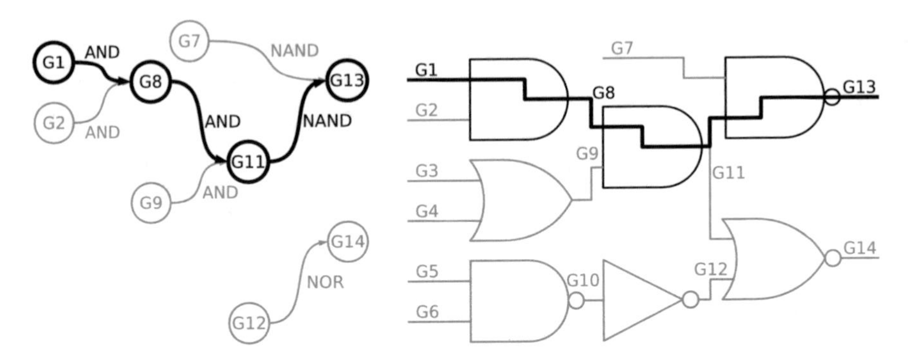

Fig. 3.8 Final graph and the original netlist showing a path that can propagate a locking value

3.3.3 Graph Analysis for Selection of Optimal Locking Nodes

At this stage, the graph is composed of several connected components. They all include at least one output, and are made of vertices that represent nodes able to propagate a locking value. These connected components can be classified in the four different categories depicted in Fig. 3.9.

In the first situation, shown in Fig. 3.9a, there is only one source vertex. Therefore, since the graph is directed, this vertex necessarily spans all the outputs, and can lock them all. It is consequently selected as the node to lock.

The second possibility, shown in Fig. 3.9b, occurs when a connected component comprises multiple source vertices but only one output. In order to embed the locking node as deeply as possible in the netlist, the distance between all source nodes and the output is computed. The furthest node from the output is selected as the node to lock.

In the case depicted in Fig. 3.9c, there are multiple source vertices too. Some source vertices, however, do not span all the outputs. In order to lock as many outputs as possible with the smallest number of nodes to be modified, only the nodes spanning all the outputs are kept. If many nodes span all the outputs, then, as previously, the furthest one from the output is selected.

In the last situation, shown in Fig. 3.9d, multiple source vertices span multiple outputs, but none spans them all. The way to proceed here is to sort the source vertices according to the number of outputs they span. Next, they are greedily selected and added to the list of nodes to lock. This process is carried out until all the outputs are locked.

Note that the situations described above are sorted according to their computational complexity. The last case, which is the most computationally expensive, is also by far the least frequent.

One we have a list of nodes to modify, the last step is to add the extra locking gates that will be responsible for forcing these nodes to a specific value if the wrong key is applied.

Fig. 3.9 **a** One source vertex, **b** Multiple source vertices, one output, **c** Multiple source vertices, multiple outputs, one (or more) source vertex spans all the outputs, **d** Multiple source vertices, multiple outputs, no vertex spanning all the outputs

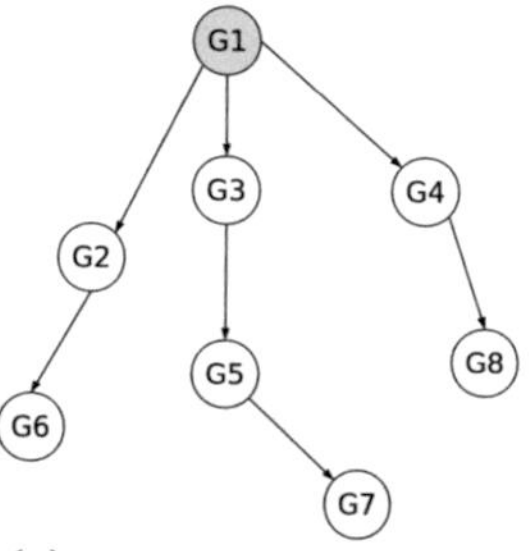

(a) One source vertex

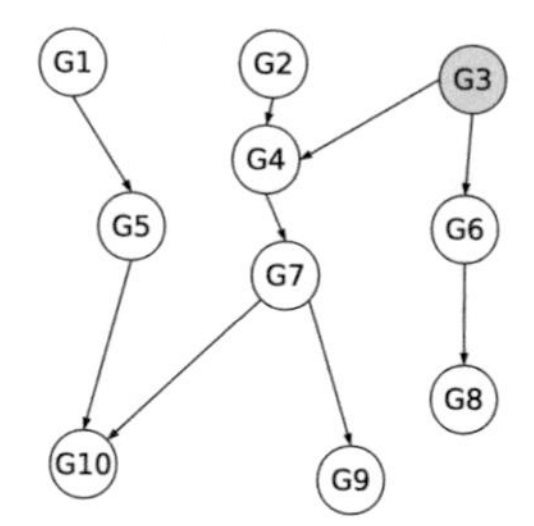

(b) Multiple source vertices, one output

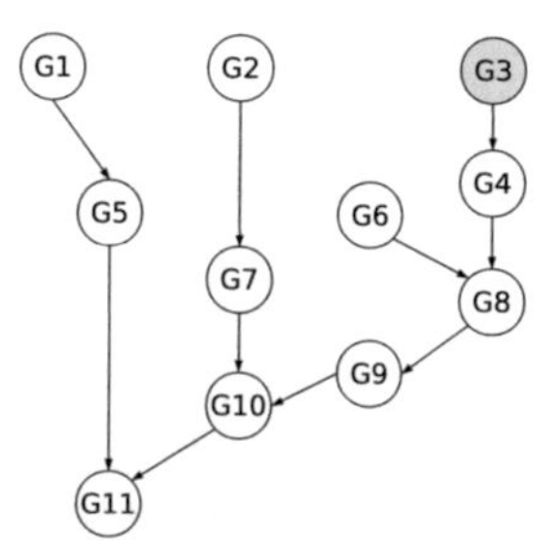

(c) Multiple source vertices, multiple outputs, one (or more) source vertex spans all the outputs

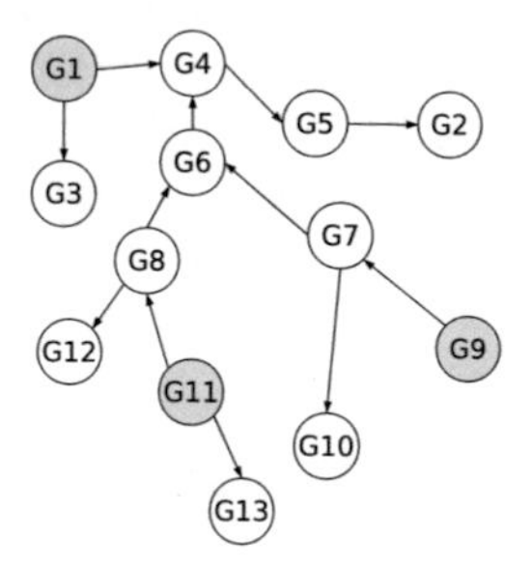

(d) Multiple source vertices, multiple outputs, no vertex spanning all the outputs

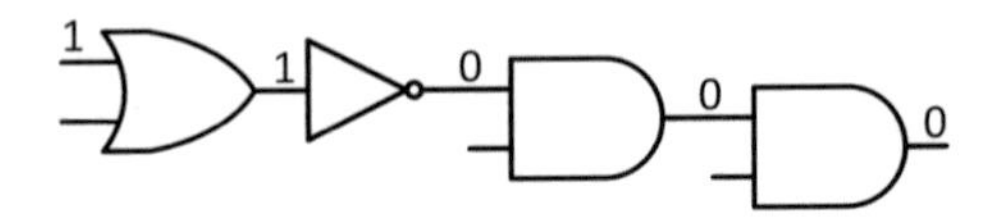

Fig. 3.10 Type of gate to insert according to the V_{forced} value and the associated unlocking bit

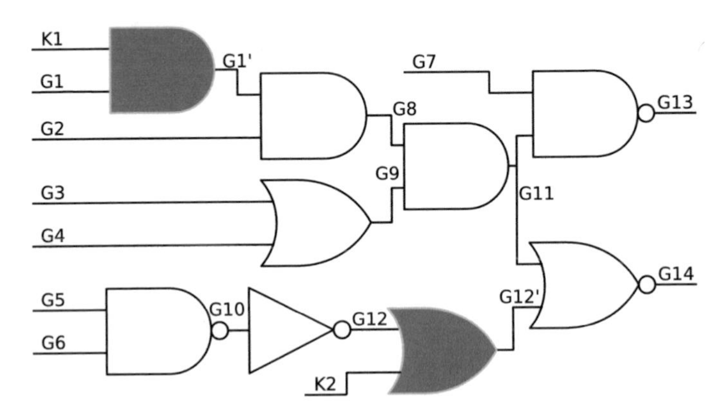

Fig. 3.11 Lockable netlist, inserted locking gates are in *gray*. The unlocking word is $(K_1K_2) =$ "10"

3.3.4 Netlist Modification

Now that we know which nodes to act on, the extra logic gates must be inserted. They will force these nodes to a specific value. The value to which each node must be forced is given by V_{locks}, the controlling value of the subsequent gate. If a node must be forced to 0, then an *and* gate is used. If a node must be forced to 1, then an *or* gate is used. This is shown in Fig. 3.10. The associated unlocking bit is the inverse of the controlling value of the inserted logic gate.

Coming back to the previous example, the nodes to be modified are G1 and G12. For G1, $V_{locks} = 0$ and for G12 $V_{locks} = 1$. Then the associated unlocking word (K_1, K_2) is "10." An *and* gate is used to force G1 to 0 if the wrong unlocking bit is applied, in this case: 0. An *or* gate is used to force G12 to 1 if the wrong unlocking bit is applied, in this case: 1. The final, lockable netlist is shown in Fig. 3.11.

3.4 Implementation Results

3.4.1 Logic Resources Overhead

The logic locking algorithm was implemented in Python, and makes use of the *igraph* module to handle graphs. We implemented the locking scheme on ITC'99

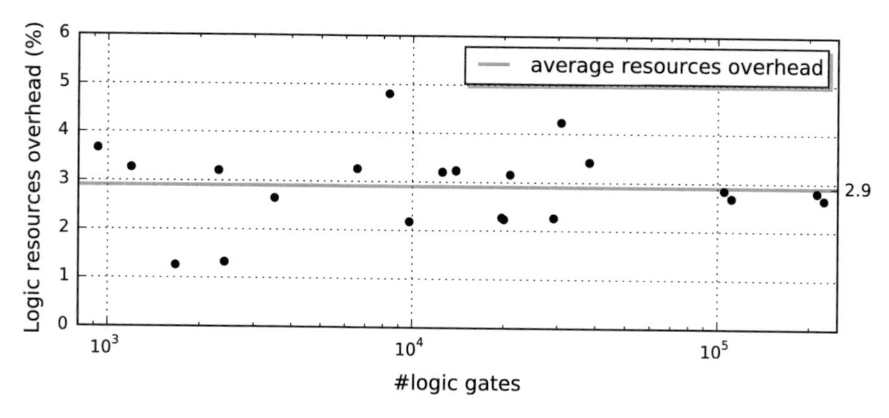

Fig. 3.12 Logic resources overhead obtained for logic locking

combinational benchmarks [19]. The netlists are described in VHDL. These benchmarks range from 1–225 k gates. The logic resources overhead is measured as the percentage of logic gates that must be added to the netlist in order to make it totally lockable. Results are shown in Fig. 3.12. The average resources overhead is 2.9 %. This is acceptable, and almost twice lower than the one authors obtained in [4]. Another interesting feature here is that the overhead remains approximately the same despite the increase in the number of gates in the original design. Protecting large netlists is consequently not more expensive than protecting smaller designs.

3.4.2 Analysis Time

Taking a step back, a major feature that will ensure the protection schemes are widely adopted is usability. It describes how easy it is for a designer to protect the IP core once it has been designed. In order to increase usability, a key point is the amount of time required to make the netlist lockable. Since these protection techniques could be integrated in EDA tools, the computation time should be reasonable. In Fig. 3.13, we provide a comparison of the computation time required to protect a netlist with both logic locking and logic masking methods. These results were obtained by executing the Python scripts on an Intel i5-4570 workstation, operating at 3.2 GHz and embedding 16Gb of RAM.

As can be seen in Fig. 3.13, the logic locking based method is more than ten thousand times faster than the method based on the logic masking. For instance, analyzing a 3,500-gate netlist requires four and a half hours with the method proposed in [4], whereas with the graph-analysis method, it takes less than one second. We extended our study to very large netlists of up to 225k gates. It turns out that the computation time increases quadratically. However, even for very large netlists, the computation

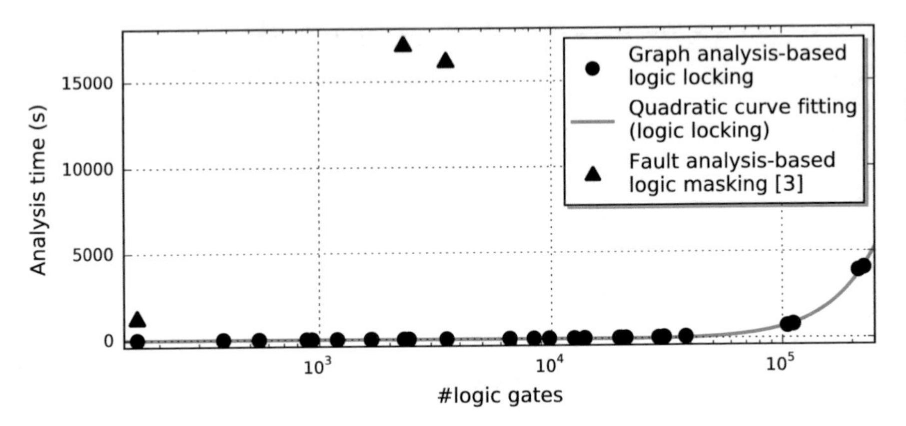

Fig. 3.13 Time required to analyze and modify the netlist

time is reasonable. For the largest one that includes 225k gates, slightly more than hour is required to make it lockable.

When it comes to the execution time, the main difference between the two protection methods is that the one proposed in [4] uses fault simulation to locate the nodes to modify. It relies on external tools that employ computationally heavy methods. Conversely, our protection technique is based on graphs, which are an effective way of representing netlists. In the context of EDA integration, our method is thus much more suitable and computationally more effective.

3.5 Evaluation

3.5.1 Correlation

In [4], the authors evaluate the efficiency of their locking scheme using the Hamming distance between the output of the original design and the output of the design when the wrong key is applied on the key inputs (i.e., when logic masking is activated). According to these authors, obtaining a 50 % Hamming distance on average is a proof that the protection scheme is efficient. However, we have shown in Sect. 3.2 that even simple circuits can exhibit such a characteristic, and that 50 % Hamming distance is simply one consequence of a zero correlation. We consequently use correlation to evaluate the efficiency of the protection scheme. The correlation is computed using Pearson's coefficient. The results are shown in Table 3.4. Since the standard deviation is zero when the outputs are locked by logic locking, Pearson's correlation coefficient is not defined. It can be considered as zero though, because when the output is locked, it provides no information about the normal behavior. Two methods are compared for logic masking: random and fault analysis-based node selection. Random selection

Table 3.4 Pearson's correlation coefficient computed for different node selection methods and key sizes

Benchmark	Key size	Logic masking		Logic locking
		Random [3]	Fault analysis [4]	Graph analysis
c432, 7 outputs, 189 nodes	32 bits	0.272	0.012	0
	64 bits	0.153	0.019	0
	128 bits	0.026	0.014	0
c5315, 123 outputs, 2362 nodes	32 bits	0.902	0.554	0
	64 bits	0.873	0.357	0
	128 bits	0.820	0.277	0
c7552, 108 outputs, 3612 nodes	32 bits	0.952	0.254	0
	64 bits	0.920	0.235	0
	128 bits	0.761	0.217	0

[3] rapidly becomes inefficient when the circuit's size increases. Randomly inserting 128 *xor* gates in a 3,612-node netlist only reduces the correlation to 0.761. Fault analysis-based logic masking is more efficient, and reduces the correlation faster as the key size increases. For large netlists, however, it fails to reduce it significantly. For example, the correlation only drops from 0.254 to 0.217 when the key size increases from 32 to 128 bits on C7552. For larger designs such as the ones considered in Sect. 3.4, the performance will probably be even worse.

We can conclude from this observation that correlation should not be used to evaluate a protection scheme. It is a cryptographic property, which should be only used in the appropriate frame. We give more details about security in Sect. 3.7. Instead of correlation, we propose a metric to evaluate protection schemes based on the insertion of extra logic gates, which is presented in the following subsection.

3.5.2 Logic Locking Metric

The intrinsic feature of a protection scheme based on the insertion of extra logic gates is altering the outputs using the extra gates. Therefore, two characteristics can be used to evaluate how effective these schemes are. The first one is: how many inputs are spanned by each extra logic gate? This is related to the amount of gates that have to be inserted to ensure total locking. If one gate locks multiple outputs, it is obviously more efficient than if multiple gates are required. The locking ratio is defined as follows:

$$Locking\ ratio = \frac{\#outputs}{\#locking\ gates} \tag{3.11}$$

Since the locking gates should be inserted as deeply as possible into the netlist, a second metric is: how far is the inserted gate from the outputs? The number of logic levels between the locking gate and the outputs is consequently also computed. The average distance between the inserted gates and the outputs is computed as the average number of logic levels on the shortest path between the inserted gates and every output that is reachable from them. The results we obtained when applying our graph-based insertion method for logic locking are presented in Table 3.5.

We can see that the number of outputs spanned by each locking gates is very close to 1. This basically means that, mostly, one logic locking gate is responsible for forcing one output. This is discussed in the following section. We can also see that the number of logic levels between the locking gates and the locked outputs is low. This could be a problem if the attacker has access to the RTL description of the design. Indeed, if the locking gates are located very close to the outputs, then the attacker can identify them easily and possibly modify the netlist to bypass the locking circuitry. This is why the locking gates need to be embedded as deeply as possible in the netlist.

Table 3.5 Evaluation of the proposed node selection technique by locking ratio and mean distance to outputs

Benchmark	#logic gates	Locking ratio	Average distance to outputs (logic levels)
c432	160	1.75	1.43
b10_C	172	1.13	1
b13_C	289	1.13	1.13
c880	383	1.63	3.39
b07_C	383	1.32	1.16
c1355	546	1.03	2
b04_C	652	1.02	1.11
b11_C	726	1.03	1.19
c1908	880	1.04	1
b05_C	927	1.82	1.52
b12_C	944	1.1	1.18
c2670	1193	1.68	2.38
c3540	1669	1.1	1.82
c5315	2307	1.68	2.07
c6288	2416	1.03	1
c7552	3512	1.16	1.5
b14_1_C	6569	1.15	1.48
b15_C	8367	1.12	1.69
b14_C	9767	1.16	1.42
b15_1_C	12543	1.12	2.06
b21_1_C	13898	1.14	1.33
b20_1_C	13899	1.14	1.32
b20_C	19682	1.15	1.36
b21_C	20027	1.14	1.29
b22_1_C	20983	1.14	1.35
b22_C	29162	1.15	1.36
b17_C	30777	1.11	1.76
b17_1_C	38116	1.11	1.97
b18_1_C	105102	1.12	1.74
b18_C	111241	1.12	1.74
	Average:	1.22	1.56

To this end, dummy logic levels can be inserted between the locking gate and the output, thereby achieving additional logic obfuscation as described in Sect. 3.2. For instance, an *or* gate can be replaced by the three gates depicted in Fig. 3.14. G is the node to be forced and K is the locking/unlocking input. Another node is picked randomly and used for the dummy logic. As depicted, the output value is

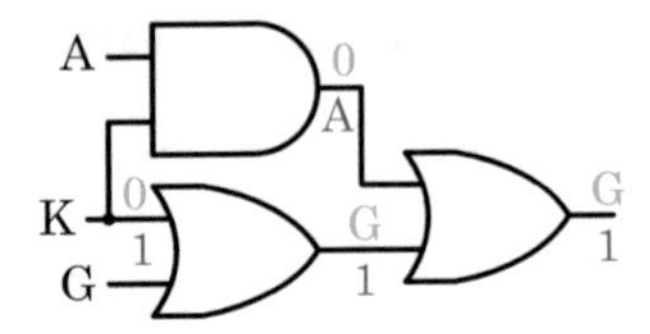

Fig. 3.14 *or* locking gate replacement with an extra logic level

either 1 or G, which means that locking is successful. Obviously, the increase in reverse engineering effort comes at the price of an increased area overhead. In order to add one logic level, three gates are inserted instead of one. If the designer wants to add a second dummy logic level, then the structure must be duplicated. Then five gates are inserted. The logic resources overhead is then $n(2k + 1)$, where n is the number of locking gates to be inserted and k is the number of dummy logic levels. In order to limit the overhead, dummy logic levels can be used only for the nodes that are too close to the outputs.

3.6 Security Analysis

3.6.1 Threat Model

To evaluate the security of logic locking, we must first distinguish the threat model of the actual context. Since we are trying to protect IP cores against illegal cloning, we must assume that the attacker has access to the original design, and can implement it. We make a stronger assumption by not limiting the number of implementations. Our aim for logic locking is only to make illegal copies nonfunctional. Thus, we first assume that the designer has access to the inputs used to unlock the design, i.e., the inputs to which the unlocking word encrypted with the secret key must be applied to unlock the circuit and to use it. In practical terms, the designer is able to write in a specific memory inside the chip, which will unlock the circuit if the correct value is provided. Moreover, since the designer appears to be legitimate at first sight, he also has access to test vectors.

3.6.2 Hill-Climbing Attack

Considering the threat model described above, a major concern expressed in [20] is the ease of a hill-climbing attack. It was described as an attack against the logic masking technique presented in [3]. However, it turns out to be equally efficient against logic locking. This is due to the tight link between the masking/locking inputs and the outputs. The attack procedure for logic masking described in [20] is as follows. First, pick a random key and apply it on the unlocking inputs. Compute the

Hamming distance between the actual and the expected output, given by the test vectors. Flip the first bit of the key. If the Hamming distance increases, then flip this bit again and repeat the action for all the bits of the key. Otherwise, if the Hamming distance decreases, move on to the next bit. The method is similar for logic locking, except that instead of using the Hamming distance as the function to minimize, the number of locked outputs is used. The main concern here is that, since there is a gradient toward the correct key in the key space, it can be easily recovered. In other words, the Hamming distance between the actual and expected output grows linearly with respect to the number of wrong key bits when logic masking is applied. Similarly, the number of outputs that are locked and the number of wrong key bits are correlated.

This is due to the fact that, as shown in Table 3.5, the ratio of the number of inserted gates to the number of outputs is close to one. In most cases, one gate is responsible for locking one output. This is a serious security concern. In this case, the security of the protection system is as low as the greatest number of key bits influencing one output. If the key bits and the outputs are connected pairwise, then the overall security level is 1 bit. In the following section, we discuss countermeasures against hill-climbing attacks.

3.6.3 A Partial Countermeasure Against Hill-Climbing Attack

In order to avoid hill-climbing attacks, the correlation between the unlocking inputs and the outputs has to be reduced. One unlocking input should have an impact on multiple outputs, in order to hide the internal relation. Similarly, every output should be locked by several key inputs.

One possible countermeasure is to add some redundancy between the locking gates and the key inputs. This can be achieved by adding inputs to the locking gates. These inputs are connected to key inputs that have the same value as the first key input of the locking gate. For example, two locking gates for which the key bit is 1 can be associated, as depicted in Fig. 3.15. It follows that in order to obtain the correct values for $G0mod$ and $G1mod$, both $K0$ and $K1$ must have the correct value. It can be extended to add more key inputs to the locking gates, and more redundancy. However, this countermeasure is only partially effective. Indeed, it only increases the equivalent security level to the number of inputs added to the locking gates. Making it secure would require the locking gates to have a very large number of inputs, which is not feasible.

After another look at the previously described characteristic, it is very similar to the diffusion property of cryptographic functions. This led us to adopt another design plan for the protection scheme. Thus the logic locking module is only responsible for disturbing the original behavior. Security is ensured by using a separate cryptographic primitive. The overall architecture is described in the following section.

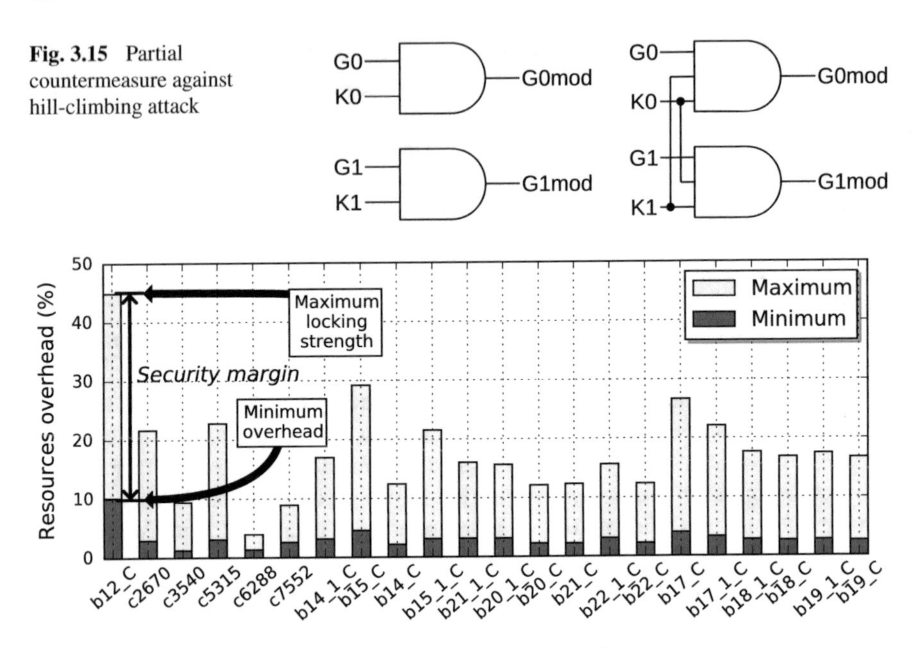

Fig. 3.15 Partial countermeasure against hill-climbing attack

Fig. 3.16 Trade-off between locking strength and resources overhead

3.7 Architecture of a Complete Design Data Protection Scheme

3.7.1 Area/locking Strength Trade-Off

Before examining the whole protection scheme architecture, let us focus on the implementation of the logic locking module. After the graph has been built and analyzed, the final graph contains nodes that are all able to propagate a locking value. The method presented in Sect. 3.3.3 to select the best nodes to modify selects as few nodes as possible in the connected components to ensure total locking, but all the other nodes are also able to lock the associated output. Therefore, some extra locking gates can be added to increase the locking strength. Indeed, if the locking signal is carried by only one wire, it could be subject to side channel attacks such as optical injection [21] and its logic value can be flipped. In fact all the nodes found in the connected components of the final graph can be modified to increase the locking strength. This comes at the cost of increased logic resources overhead. This design trade-off is illustrated in Fig. 3.16, where the logic resources related to minimum overhead and maximum locking strength are given for all ISCAS'85 benchmarks. For b15_C for instance, the minimum overhead to achieve total functional locking is 4.52 %. However, up to 29 % extra resources can be added to further strengthen logic

(a) Original netlist and modified netlist with lowest area overhead

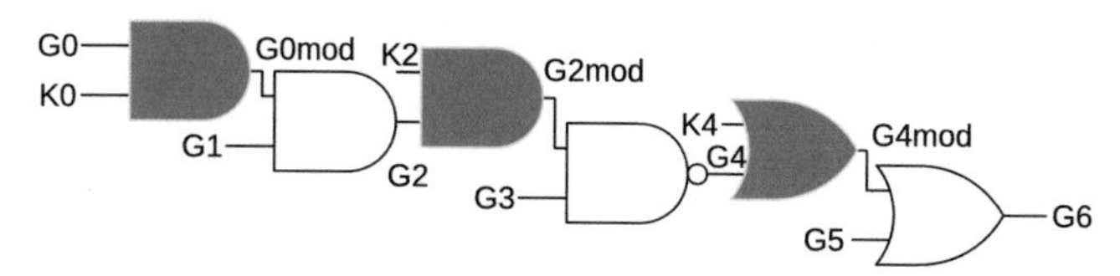

(b) Modified netlist with maximum locking strength and separated key bits

Fig. 3.17 **a** Original netlist and modified netlist with lowest area overhead. **b** Modified netlist with maximum locking strength and separated key bits

locking. The designer can decide on the acceptable resources overhead and increase the associated locking strength accordingly.

An example is given in Fig. 3.17. The original netlist and the netlist modified for logic locking with minimal overhead are shown in Fig. 3.17a. There is only one node forced, and one unlocking bit input. On the other hand, since all nodes *G*0, *G*2 and *G*4 can propagate a locking value, they can all be forced to increase the locking strength. This is shown in Fig. 3.17b. Three locking gates are inserted. The associated unlocking bits must all be set to their correct value in order to get the correct output. Of course, it comes at the price of an increased area overhead.

3.7.2 On the Need for a Cryptographic Primitive

In [4], the authors claim to achieve security by reaching 50 % Hamming distance between the original and masked outputs. Since in this case, security is not based on a cryptographic primitive, it is easily broken and [20] showed how it was possible to recover the key using a basic hill-climbing attack.

Only the system integrators allowed by the designer to unlock the IP core should be able to do so. If provable security is necessary, there is no other way than using a cryptographic primitive to obtain it. Another advantage is that such primitives, if chosen carefully, have been subject to a variety of attacks. Therefore, their security has been tested. The designer can then pick a strong cryptographic primitive that has successfully resisted multiple attacks, and implement it carefully. This will provide provable security of access to the normal operation of the IP core. For that reason, using a cryptographic primitive is necessary.

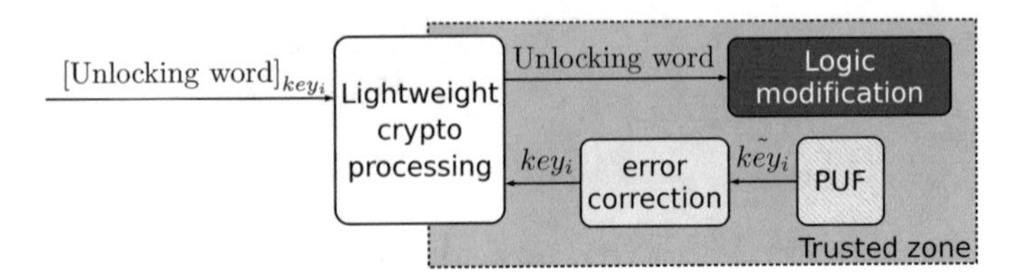

Fig. 3.18 Architecture of the proposed design protection module

3.7.3 Architecture

Owing to such considerations, we are now able to define the general architecture of the design protection scheme. It is depicted in Fig. 3.18.

The first block is the cryptographic primitive, which ensures secure access and avoids simple attacks. Using a lightweight, hardware-oriented algorithm is a good option here to limit the area overhead. The second block is a PUF, acting as a unique identifier, which is necessary in the case of IP distribution to uniquely identify all the instances of a particular design. It allows the designer to have a database containing all the IP core instances and their associated key. This is used to derive the key encrypting the unlocking word. In this way, it helps fulfill the following requirement: owning the key for one instance of the design should not help in unlocking another instance. Different types of PUFs are available, such as the TERO-PUF [22], the butterfly PUF [23] or the arbiter PUF [24]. It could also be achieved in the form of a secret word stored in nonvolatile memory. An error-correction module corrects the PUF's response. Finally, the unlocking word is deciphered and sent to the locking module. The locking module can implement logic encryption, masking or locking. Its role is to make the circuit unusable if the message sent to the cryptographic primitive is not the right unlocking word encrypted with the correct key associated with the circuit.

3.8 Summary

Design data protection schemes modifying the logic are a powerful way to render the circuit harder to reverse-engineer or unusable if it has been counterfeited. Several techniques to modify the logic are available, namely logic encryption, obfuscation, masking, or locking. They act on specific sites of the combinational logic part of the design. The method to select the sites to act on must be computationally efficient to be easily used, but also select the best sites. A graph analysis-based method is presented, which is fast and effective. Finally, we present design considerations, which include the integration of the logic modification in a wider protection scheme, in order to provide cryptographic strength and per-device uniqueness.

References

1. Y. Alkabani, F. Koushanfar, Active hardware metering for intellectual property protection and security, in *USENIX security*, USA, Boston, MA, Aug 2007, pp. 291–306
2. I. McLoughlin, Reverse engineering of embedded consumer electronic systems, in *IEEE 15th international symposium on consumer electronics*, Singapore, Singapore, June 2011, pp. 352–356
3. J.A. Roy, F. Koushanfar, I. Markov, EPIC: ending piracy of integrated circuits, in *Design, automation and test in Europe* (2008), pp. 1069–1074
4. J. Rajendran, H. Zhang, C. Zhang, G.S. Rose, Y. Pino, O. Sinanoglu, R. Karri, Fault analysis-based logic encryption. IEEE Trans. Comput. **64**(2), 410–424 (2015)
5. J. Rajendran, Y. Pino, O. Sinanoglu, R. Karri, Logic encryption: a fault analysis perspective, in *Design, automation & test in Europe conference*, Dresden, Germany, March 2012, pp. 953–958
6. S. Dupuis, P. Ba, G. Di Natale, M. Flottes, B. Rouzeyre, A novel hardware logic encryption technique for thwarting illegal overproduction and hardware trojans, in *IEEE International On-Line Testing Symposium*, Girona, Spain, Platja d'Aro, June 2014, pp. 49–54
7. G. Hachez, A comparative study of software protection tools suited for e-commerce with contributions to software watermarking and smart cards, Ph.D. dissertation, Université Catholique de Louvain, March 2003
8. J. Rajendran, Y. Pino, O. Sinanoglu, R. Karri, Security analysis of logic obfuscation, in *Annual design automation conference*, San Francisco CA, USA, June 2012, pp. 83–89
9. R.S. Chakraborty, S. Bhunia, HARPOON: an obfuscation-based SoC design methodology for hardware protection. IEEE Trans. Comput.-Aid. Des. Integr. Circ. Syst. **28**(10), 1493–1502 (2009)
10. R. Torrance, D. James, The state-of-the-art in semiconductor reverse engineering, in *Proceedings of the design automation conference, DAC 2011*, San Diego, California, USA, 5–10 Jun 2011, 2011, pp. 333–338
11. M. Brzozowski, V.N. Yarmolik, Obfuscation as intellectual rights protection in VHDL language, in *International conference on computer information systems and industrial management applications*, IEEE Computer Society, Elk, Poland, Jun 2007, pp. 337–340
12. U. Meyer-Baese, E. Castillo, G. Botella, L. Parrilla, A. Garca, Intellectual property protection (IPP) using obfuscation in C, VHDL, and verilog coding, in *SPIE defense, security, and sensing*, Orlando, Florida, USA, Jun 2011
13. J. Rajendran, M. Sam, O. Sinanoglu, R. Karri, Security analysis of integrated circuit camouflaging, in *ACM conference on computer & communications security*, Berlin, Germany, Nov 2013, pp. 709–720
14. SypherMedia, Circuit camouflage technology (2012)
15. A. Baumgarten, A. Tyagi, J. Zambreno, Preventing IC piracy using reconfigurable logic barriers. IEEE Des. Test Comput. **27**(1), 66–75 (2010)
16. E. Jung, C. Hung, M. Yang, S. Choi, An locking and unlocking primitive function of FSM-modeled sequential systems based on extracting logical property. Int. J. Inf. **16**(8), 6279–6290 (2013)
17. M.T. Rahman, D. Forte, Q. Shi, G.K. Contreras, M.M. Tehranipoor, CSST: preventing distribution of unlicensed and rejected ICs by untrusted foundry and assembly, in *IEEE international symposium on defect and fault tolerance in VLSI and nanotechnology systems*, Netherlands, Amsterdam, Oct 2014, pp. 46–51
18. A. Basak, Y. Zheng, S. Bhunia, Active defense against counterfeiting attacks through robust antifuse-based on-chip locks, in *IEEE 32nd VLSI test symposium*, USA, Napa CA, Apr 2014, pp. 1–6
19. S. Davidson, ITC'99 benchmark circuits—preliminary results, in *IEEE international test conference*, NJ, USA, Atlantic City, Sept 1999, p. 1125
20. S.M. Plaza, I.L. Markov, Protecting integrated circuits from piracy with test-aware logic locking, in *International conference on computer aided design*, San Jose, CA, USA, Nov 2014

21. S.P. Skorobogatov, R.J. Anderson, Optical fault induction attacks, in *International workshop on cryptographic hardware and embedded systems*, San Fransisco CA, USA, Aug 2002
22. L. Bossuet, X.T. Ngo, Z. Cherif, V. Fischer, A PUF based on transient effect ring oscillator and insensitive to locking phenomenon. IEEE Trans. Emerg. Top. Comput. **2**(1), 30–36 (2014)
23. S.S. Kumar, J. Guajardo, R. Maes, G.J. Schrijen, P. Tuyls, The butterfly PUF protecting IP on every FPGA, in *IEEE international workshop on hardware-oriented security and trust*, USA, Anaheim CA, Jun 2008, pp. 67–70
24. J.W. Lee, D. Lim, B. Gassend, G.E. Suh, M. van Disk, S. Devadas, A technique to build a secret key in integrated circuits for identification and authentication applications, in *Symposium on VLSI circuits*, Jun 2004, pp. 176–179

Chapter 4
IP FSM Watermarking

Edward Jung and Lilian Bossuet

4.1 Introduction

Hardware design reuse has been the viable solution to deal with the ever-increasing logic density in the semiconductor industry. For instance, an application-specific integrated circuit (ASIC) architecture can be designed using previously designed subcomponents or subsystems. Hardware design intellectual property (IP) is a design unit that can be viewed as an independent subcomponent of a complete design (e.g., SOC design.) Examples of design unit include abstract algorithm, technique, or methodology that can make the design better as well as physical design blocks such as embedded controllers and fully routed *netlist*. This can result in new products to be on the market in time and at a cost-effective way. The newly developed subcomponents can also be tested and deposited as new design IPs in the IP library for future reuse.

Despite the attractiveness of reuse-based design, IP owners and vendors have encountered IP piracy and infringement. In practice, most design IPs are secured by deterrent and protection mechanisms. Deterrent protections are usually provided by patents, copyrights, contracts, trademarks, and trade secrets. But this only discourages the misuse of IPs. Protection mechanisms are more proactive than the deterrent solution. It usually uses technical means such as cryptographic algorithms, dedicated hardware, or even chemicals to prevent unauthorized access to the IP. Protecting design IPs using encryption and other protection mechanisms make

E. Jung (✉)
Computer Science Department, Kennesaw State University,
Marietta, GA 30060, USA
e-mail: ejung4@kennesaw.edu

L. Bossuet
Laboratoire Hubert Curien, CNRS UMR 5516, Université Jean Monnet,
42000 Saint-Etienne, France
e-mail: lilian.bossuet@univ-st-etienne.fr

L. Bossuet and L. Torres (eds.), *Foundations of Hardware IP Protection*,
DOI 10.1007/978-3-319-50380-6_4

IP piracy more difficult and more expense. However, detection schemes such as a watermark can enable IP owner to identify the occurrence of IP piracy. Both the hardware design reuse and the IP piracy and infringement problems are well versed in [1].

As a solution to the problem of protecting hardware IPs, one of the key initiatives known as *Salware* was proposed [2]. Also, a variety of techniques at various levels and/or at different stages of the design process have been proposed. In particular, many interesting methods have been developed for solving IP protection problems: graph partitioning [3], constraint-based optimization [4], graph coloring [5], information hiding [6], FPGA mapping and designs [7], and high-level behavioral optimization [8, 9].

Controllers are an important component in designing an embedded hardware system. The design of a controller can be specified using a synchronous finite-state machine (FSM). The purpose of this chapter is to introduce the basic concepts of intellectual property (IP) watermarking and approach to secure design IPs of control units modeled in a synchronous finite-state machine (FSM) from unauthorized use. In general, an FSM watermarking consists of two processes: an *embedding* process and a *verification* process. We focus on the embedding process in this chapter.

4.1.1 Basic Concepts of IP Watermarking

The generic IP watermarking scheme is shown in Fig. 4.1, similar to the model [10]. It consists of the embedding and detection (i.e., verification) processes. The embedding process is a mapping of the form $O \times K \times W \rightarrow \widetilde{O}$, where the object to be protected is denoted by O, a watermark by W, and a watermark key by K. Its output $\widetilde{O}$ is a watermarked object. The detection process produces as an output either the recovered watermark W or some kind of confidence measure C indicating how likely an object-under-test $\widetilde{O}'$ is distinguishable from the watermarked object $\widetilde{O}$. An object can be in a variety of forms. For instance, an object may include the creative design of digital systems or ICs such as a sequential logic component, which can be modeled as a finite-state machine (FSM).

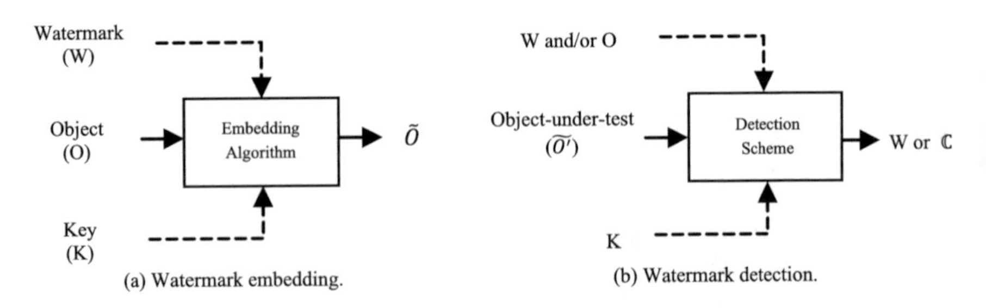

Fig. 4.1 Generic IP watermarking process

4.1.2 Types of Watermarking System

In general, we can classify the types of FSM IP watermarking systems into five categories as below [10]:

- *Distinguishing* $\widetilde{M'}$ *from* $\widetilde{M}$ (Type I): a *Type I* system uses both a copy of the watermarked FSM device and the original watermarked device, and yields a "yes" or "no" answer to the question: *is* $\widetilde{M'}$ *distinguishable from* $\widetilde{M}$*?* ($\widetilde{M'} \times \widetilde{M} \times K \times W \to \{0, 1\}$). $\widetilde{M'}$ is a device-under-test and $\widetilde{M}$ is the original watermarked device.
- *Containment of W in* $\widetilde{M'}$ (Type II): a *Type II* system does not use $\widetilde{M}$ for detection but answers to the question*: does* $\widetilde{M'}$ *contain the watermark W?* ($\widetilde{M'} \times K \times W \to \{0, 1\}$).
- *Extraction of W from* $\widetilde{M'}$ (Type III): *Type III* systems aim to find *where* the watermark W could be in $\widetilde{M'}$. It requires extracting *W* from the possibly illegal copy $\widetilde{M'}$ and may use $\widetilde{M}$ for detection.

The types of systems above can be used to prove the ownership of IP in court and control the copying of devices. Many of the currently proposed schemes fall in this category.

There are two other types of systems we can consider under special circumstances. One such system is modeled as $\widetilde{M'} \times K \to W$, where the exact information about the watermark can be extracted from $\widetilde{M'}$ without the availability of $\widetilde{M}$. This problem may be the most challenging and it is unknown if such a solution exists. Finally, we can consider a system that allows any user to read the watermark *W* without being able to remove it. In this chapter, the model we follow is ***a Type I system***.

4.1.3 FSM IP Watermarking: State of the Arts

In this section, we focus FSM IP watermarking and review the related works. For other types of IP protection approaches, refer to the survey paper [11].

In the early days of FSM IP protection, hiding a secret watermark in a sequential circuit was first proposed in [12, 13]. In these works, the watermarking was performed by modifying the State Transition Graph (STG) to go through a chosen path of state transitions with a certain set of inputs (i.e., signatures). The watermark design was done in such a way that the insertion of the watermark did not have any effect on the IC's functionality. The proof of IP ownership was ensured by the fact that the displayed input-transition behavior would be extremely rare in a nonwatermarked circuit. In another work [14], the FSM IP watermarking was based on extracting the unused transitions in the STG, while extra transitions were added to satisfy the design goals.

Later, the different idea of hiding information in the unused transitions of FSM was proposed [6]. They developed a SAT-based algorithm to find the maximal set of redundant transitions for a given minimized FSM and utilized this redundancy to hide the information in the FSM without changing the given minimized FSM. Another interesting solution was proposed [15]. In this work, multiple watermarks were added to further enhance security and it was shown that hiding multiple watermarks in STG is an instance of obfuscating a multiple point function with a general output. The basic idea of this work was to integrate the user-defined watermarked FSMs (i.e., the designer's secret) into the original FSM.

More recently, an FSM IP watermarking scheme by making the authorship information a nonredundant property of the FSM was proposed [16]. In this work, the watermark bits were interwoven into the outputs of the existing and free transitions of STG. Pseudo input variables have been reduced and made functionally indiscernible by the notion of reversed free literals. Then, the reserved literals were assigned to minimize the overhead of watermarking and made the watermarked FSM fallible upon removal of any pseudo input variable. They showed the lower or acceptably low overheads with higher tamper resilience and stronger authorship proof in comparison with related watermarking schemes. Another method was proposed [17]. In this work, a set of edges between states were added as a dummy entity. This was done by controlling state encoding values. The new edges created by this method were paired with an unused state input combination, and the output was specified as a *don't-care* condition.

These FSM IP watermarking schemes can be modeled as shown in Fig. 4.2. In the embedding process, the original FSM denoted by M is transformed into M^+ which contains redundant information (i.e., new dummy or nonfunctional states and/or state transitions.) In the verification process, it usually checks how likely a device-under-test $\widetilde{M'}$ contains the watermark, or it is distinguishable from the watermarked device, indicated by confidence measure C. However, these state-level solutions require *accessing* internal states (i.e., flip-flop values) using verification component V (e.g., a partial or full scan chain.).

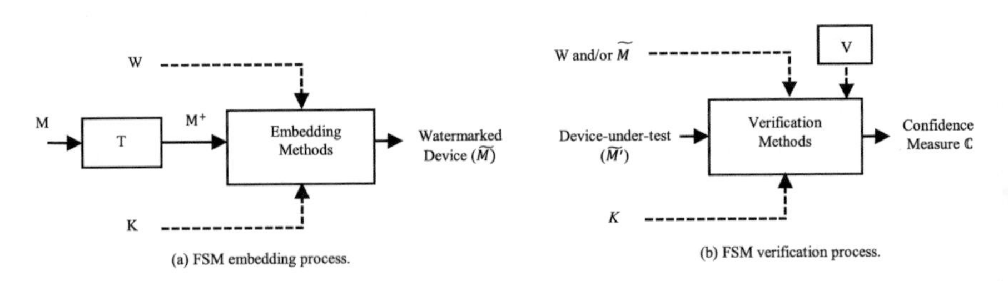

Fig. 4.2 FSM IP watermarking scheme (with redundancy)

In general, these FSM IP watermarking methods can be effective as demonstrated in those papers, especially when the overhead due to the addition of watermarked data is minimally introduced. However, these redundancy-based methods have the common limitations of augmenting the original FSM with additional states and/or state transitions.

4.1.4 Design Philosophy

As briefly mentioned in the previous section, most of the hardware FSM IP protection solutions have been developed by hiding secret information (e.g., watermarking) that is *intentionally* added to the circuit in order to prevent illegal use of the IC. The value of an IP protection solution has been determined by its efficiency of implementation in terms of reducing the design overheads (e.g., area, delay, power). For instance, some *nonfunctional* states can be added to the original FSM in order to insert watermarks. In the normal mode of operation, only functional states are accessed, while the added states (i.e., nonfunctional) are allowed to be accessed during the watermarking operation. Most of the previous work aimed to add nonfunctional entities, made as small as possible, to reduce the overheads while trying to satisfy the watermark design goals such as achieving high robustness or low coincidence (i.e., a collision) at the same time. Another key characteristic of the previous work on FSM watermarking is to define a watermark *at a state level*. For instance, a watermark can be defined in terms of a set of visited states (usually nonfunctional states), which are traveled upon by applying specific input signals known as signatures. In an analogy, this type of approach would be similar to adding tattoos (either big or small) on the external surface of the human skin to hide the designer's secret information.

We challenge this traditional design philosophy by raising the following question. Then, we describe a new method.

- *Would it be feasible to both construct (i.e., embed) and detect a watermark at an FSM level without adding nonfunctional entities (e.g., states, state transitions)?*

Here, an "FSM-level" watermark refers to the designer's secret information that is stored within an FSM itself, instead of an ordered set of individual states. The advantage of this is twofold: (1) There is no addition of nonfunctional entities during the embedding process; (2) There is no need to check individual states during the detection process. This type of approach would be similar to extracting part of a DNA sequence *within* the body of a person and using it for the designer's secret information.

4.1.5 FSM IP Watermarking: Nonredundancy-Based Approach

Nonredundancy-based approach is schematically shown in Fig. 4.3. The watermarked FSM $(\widetilde{M})$ is a function of some inherent property within the FSM M itself. The verification is performed at an FSM level without accessing individual/internal states (i.e., flip-flops). Confidence measure C indicates how likely the device-under-test $\left(\widetilde{M'}\right)$ *is distinguishable from the watermarked FSM* $(\widetilde{M})$?

In the rest of this chapter, we describe one of nonredundancy-based embedding methods using a hierarchical state encoding.

4.2 Problem Formulation and Solution Architecture

In this section, basic definitions for the FSM system model are reviewed. Then, research problems are formulated, followed by solution architecture. Finally, a motivational example is provided.

4.2.1 Basic Definitions

We provide the basic definitions [18, 19] that will be used throughout this chapter.

Definition 4.1 A *synchronous finite-state machine* (or *sequential machine*) is a quintuple $M = (S, I, O, \delta, \lambda)$, where (i) S is finite nonempty set of states; (ii) I is a finite nonempty set of inputs; (iii) O is a finite nonempty set of outputs; (iv) $\delta: S \times I \rightarrow S$ is the transition function; (v) $\lambda: S \times I \rightarrow O(Mealy\, type)$, $\lambda: S \rightarrow O(Moore\, type)$.

A Mealy-type machine can be converted into a Moore-type machine and vice versa. Unless it is needed to distinguish one type from the other, Mealy-type machine is assumed in this chapter. Synchronous finite-state machine (FSM) or sequential machine (M) will be used interchangeably.

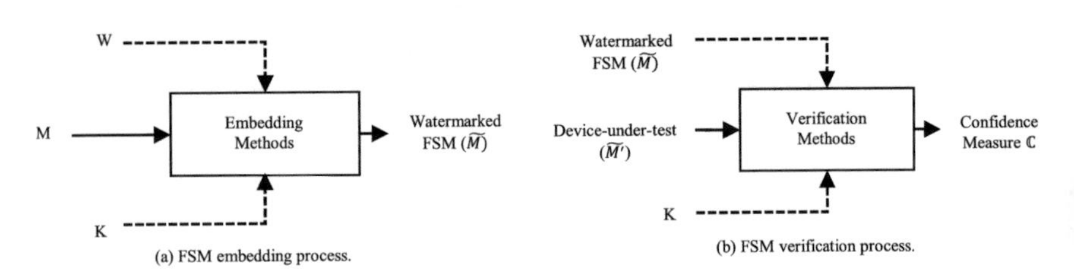

Fig. 4.3 FSM IP watermarking scheme (without redundancy)

Definition 4.2 A *state machine* is a triplet $M = (S, I, \delta)$, where (i) S is a finite nonempty set of states; (ii) I is a finite nonempty set of inputs; (iii) $\delta: S \times I \rightarrow S$ is a transition function.

In the following, we define a general partition on the states of the machine (i.e., *state partition*), and the ordering relation of state partitions.

Definition 4.3 A *partition* φ on S of the machine $M = (S, I, O, \delta, \lambda)$ is a collection of disjoint subsets of S whose set union is S. That is, $\varphi = \{B_\alpha\}$ such that $B_\alpha \bigcap B_\beta = \varnothing$ for $\alpha \neq \beta$. and $\bigcup\{B_\alpha\} = S$..

We refer to the sets of φ as *blocks* of φ. and designate the block containing s by $B_\varphi(s)$. In writing out a partition, we distinguish blocks with bars and semicolons.

Example 4.1 If $S = \{1, 2, 3, 4, 5, 6, 7, 8\}$ and partition φ on S has blocks $\{1, 3, 4, 5\}$, $\{2, 6\}$, and $\{7, 8\}$, then we write $\varphi = \{\overline{1, 3, 4, 5}; \overline{2, 6}; \overline{7, 8}\}$.

Note that φ is a set and has elements like any other set. For φ_1 and φ_2 on S, we say that φ_2 is *larger than or equal to* φ_1, and write $\varphi_1 \leq \varphi_2$, if and only if every block of φ_1 is contained in a block of φ_2.

Example 4.2 If $\varphi_1 = \{\overline{1, 2}; \overline{3, 4}; \overline{5, 6}; \overline{7, 8}\}$ and $\varphi_2 = \{\overline{1, 2, 3, 4}; \overline{5, 6, 7, 8}\}$, then $\varphi_1 \leq \varphi_2$.

Definition 4.4 Two partitions φ_1 and φ_2 are *equal*, $\varphi_1 = \varphi_2$, if and only if $\varphi_1 \leq \varphi_2$ and $\varphi_2 \leq \varphi_1$.

We write $s \equiv t(\varphi)$ if and only if s and t are contained in the same block of φ. That is, $s \equiv t(\varphi)$ if and only if $B_\varphi(s) = B_\varphi(t)$.

Definition 4.5 A partition φ on the set of states of $M = (S, I, O, \delta, \lambda)$ is a *closed* partition if and only if $\mathrm{s} \equiv \mathrm{t}(\varphi)$ implies that $\delta(\mathrm{s}, \mathrm{a}) \equiv \delta(\mathrm{t}, \mathrm{a})(\varphi)$ for all a in I.

Now, we can define a "multiplication" operation (denoted by) on partitions on a set.

Definition 4.6 If φ_1 and φ_2 are partitions on S, then $\varphi_1 \cdot \varphi_2$ is the partition on S such that $s \equiv t(\varphi_1 \cdot \varphi_2)$ if and only if $s \equiv t(\varphi_1)$ and $s \equiv t(\varphi_2)$.

Example 4.3 If $\varphi_1 = \{\overline{1, 2}; \overline{3, 4}; \overline{5, 6}; \overline{7, 8, 9}\}$ and $\varphi_2 = \{\overline{1, 6}; \overline{2, 3}; \overline{4, 5}; \overline{7, 8}; \overline{9}\}$, then $\varphi_1 \cdot \varphi_2 = \{\overline{1}; \overline{2}; \overline{3}; \overline{4}; \overline{5}; \overline{6}; \overline{7, 8}; \overline{9}\}$.

Definition 4.7 If partition φ_1 on S has the same singleton element as S, then the partition φ_1 is called a "*zero*" partition (denoted by $\varnothing$). If partition φ_2 on S has one block containing all elements in S, then the partition φ_2 is called an "*identity*" partition.

Definition 4.8 If partition φ on S is either "zero" or "identity" partition, then the partition φ is called a "*trivial*" partition. Otherwise, it is called a "*nontrivial*" partition.

Example 4.4 If $S = \{1, 2, 3, 4, 5, 6, 7, 8\}$ and partition $\varphi_1 = \{\overline{1}; \overline{2}; \overline{3}; \overline{4}; \overline{5}; \overline{6}; \overline{7}; \overline{8}\}$, then $\varphi_1 = \varnothing$. If $S = \{1, 2, 3, 4, 5, 6, 7, 8\}$ and partition

$\varphi_2 = \{\overline{1,2,3,4,5,6,7,8}\}$, then φ_2 is an identity partition. Both φ_1 and φ_2 are "trivial." Both φ_1 and φ_2 in Example 4.3 are "nontrivial."

Definition 4.9 A partition $\varphi = \{\overline{B_1}; \overline{B_2}; \ldots; \overline{B_t}\}$ on S is an *input-consistent (or input-independent) partition* if $\delta(\overline{B}_i, a) = \delta(\overline{B}_i, b)$ for a, b in I, and $i = 1, 2, \ldots. t$.

That is, the state behavior of a component with an input-consistent partition is independent of input I. Note that state encoding is done based on partitioning states into a set of blocks. For instance, if $S = \{1, 2, 3, 4, 5, 6, 7, 8\}$ and partition φ on S has three blocks $\{1, 3, 4, 5\}$, $\{2, 6\}$, and $\{7, 8\}$, i.e., $\varphi = \{\overline{1,3,4,5}; \overline{2,6}; \overline{7,8}\} = \{B_1; B_2; B_3\}$, then one possible way of assigning binary values to each state in a unique way, known as state assignment or state encoding (which can be denoted by e_1), is $\{(B_1, 00), (B_2, 01), (B_3, 10)\}$. Note, however, that the relation between the state partition and the state encoding is *not* a one-to-one and onto function, since other state encoding $e_1^{'}(\neq e_1)$ can be made for the same state partition. For instance, $e_1^{'} = \{(B_1, 00), (B_2, 11), (B_3, 10)\} \neq e_1$ for the same state partition φ.

4.2.2 Research Problems

We are interested in using the FSM state encoding scheme to embed a watermark which possesses some sequential property (i.e., a cyclic behavior in this work). We envision of creating a watermark at an FSM level. In particular, two specific questions are: (1) *can a hierarchical state encoding be used to embed a watermarked FSM* ($\widetilde{M}$)? (2) *If so, what are the necessary and sufficient conditions for the existence of such embedding solution?*

The uniqueness of a watermarked FSM should depend on a chosen sequential property. In illustrating the basic ideas of the proposed embedding approach, we consider the sequential property of *cyclic behavior* with the two additional conditions: (1) *maximal periodicity* and (2) *input consistency*. The chosen sequential property will be denoted by P^* as a general term, but by a specific notation of p_i when other selection options are available from a set of $\{p_1, p_2, \ldots, p_s\}$. Note that the property P^* is a *strong* and composite property. The rationale for choosing this strong property is twofold. First, if P^* exists in an FSM, then an effective watermarking scheme can be designed. Second, P^* is one of the common properties in any embedded controller or sequential FSM designs. For instance, a "*t-bit*" counter is typically present in an embedded controller. In general, there is a trade-off between choosing a strong versus weak property in achieving desirable IP watermarking solutions.

4.2.3 Solution Architecture

The general solution architecture, shown in Fig. 4.4, is based on the decomposition of an FSM. It has two FSM subcomponents, FSM_w and FSM_r, where FSM_w is a watermarked FSM while FSM_r is a residual FSM. The residual FSM is needed to preserve the original functionality of FSM (i.e., $FSM = FSM_w \oplus FSM_r$, where $\oplus$ indicates the concurrent operation). Note that both FSM_w and FSM_r are state machines themselves which include the state (S_w, S_r), the state transition (δ_w, δ_r), and output functions (λ_w, λ_r). It can be shown that a given FSM can be always decomposed into two independent FSMs, i.e., the cascaded/serial decomposition [19].

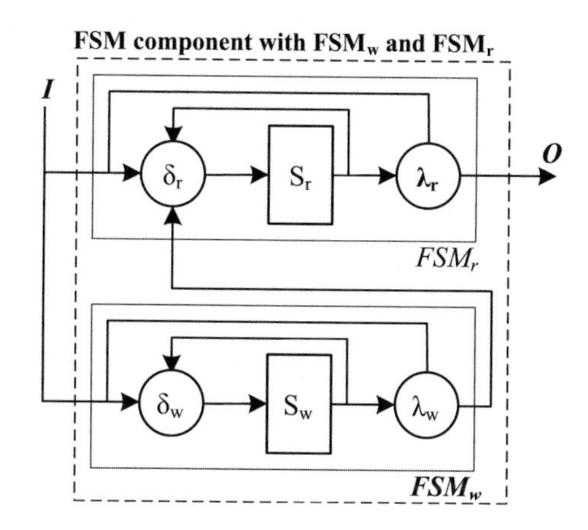

Fig. 4.4 General solution architecture

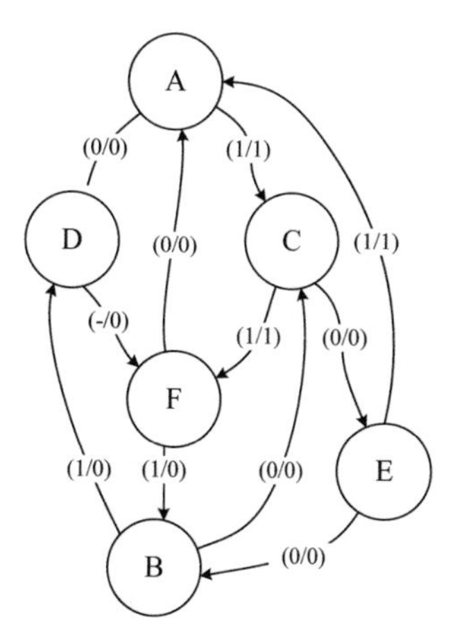

Fig. 4.5 An example FSM M

4.2.4 Motivational Example

To illustrate the basic ideas of the proposed embedding method, we use the FSM M as shown in Fig. 4.5 [19]. Note that (x/z) indicates an input x and an output z, and "_" denotes a *don't-care* condition.

The M has the six states {A, B, C, D, E, F}, input x, and output z. The state transition and the output function are described in the diagram. Figure 4.6 shows the decomposition of M into two submachines. Here, from the FSM IP watermarking perspective, the work of decomposition can be interpreted as the extraction of M_p from M where M_p shows a cycle of three states: α, β, γ. The successor M_s is to preserve the original functionality of M. This is based on a cascade decomposition. Note M_p possesses a cyclic state behavior. We propose to use the extracted M_p as the watermarked FSM.

Using the machine decomposition and state encoding methods (i.e., state assignment problem) [19], the following state encoding can extract M_p from M: e_1 = {(A, 000), (B, 001), (C, 010), (D, 011), (E, 100), (F, 101)} with {(α, 00), (β, 01), (γ, 10)} and {(a, 0), (b, 1)}. Note that the original states are realized by two internal states: A = (α, a), B = (α, b), C = (β, a), D = (β, b), E = (γ, a), F = (γ, b). For instance, the state "A" can be realized by two internal states "α" and "a" using three flip-flops, if the minimum number of flip-flops is used.

Note that the state encoding above *uniquely* combines the following two states together: A and B, C and D, and E and F, producing a three-block partition $\pi = \{\alpha; \beta; \gamma\} = \{\overline{A,B}; \overline{C,D}; \overline{E,F}\}$ for $M_p(\equiv FSM_w)$ as shown in Fig. 4.7a. No other two-state combinations (e.g., A and C) except this unique partition π will generate this special property. For instance, if the partition is made with $\pi' = \{\overline{A,C}; \overline{B,D}; \overline{E,F}\}$, as shown in Fig. 4.7b, no such property can be generated.

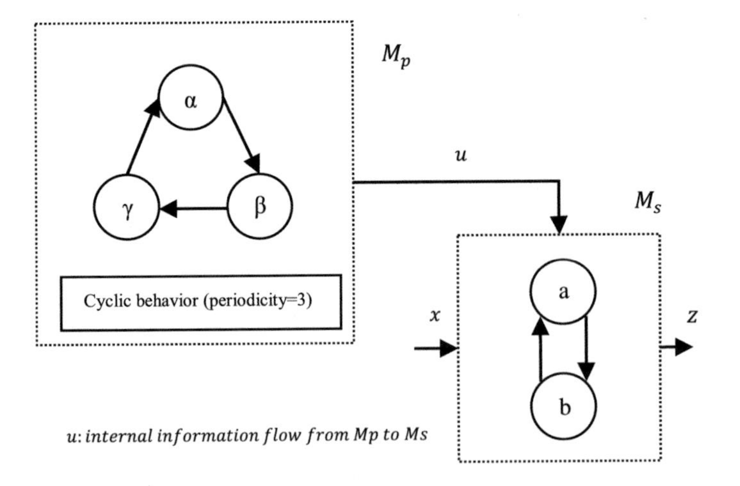

Fig. 4.6 Extraction of M_p from M

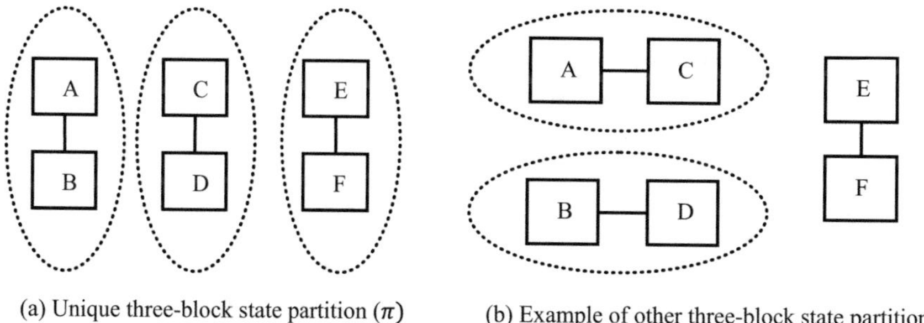

Fig. 4.7 Uniqueness of the state partition (π) generating a specific property P*

4.3 The Embedding Method

4.3.1 Hierarchical State Encoding

The central ideas of the embedding method lie in (1) *extracting* a watermarked FSM (FSM_w) possessing a sequential property (p_j), and (2) *performing* a specific state encoding (e_i). The conceptual diagram is shown in Fig. 4.8a. For completeness, the verification is shown in Fig. 4.8b.

State Encoding: For a given *n*-state FSM and *m*, the number of variables (or flip-flops), there exists potentially many different state encodings. Let Ω be the set of all possible encodings: $\Omega = \{e_1, e_2, \ldots, e_t\}$. The encoding space Ω can be quite large. In general, the number of flip-flops that can be used is a variable. It makes a direct impact on the size of the encoding space (i.e., $|\Omega| = t$). One possible case is

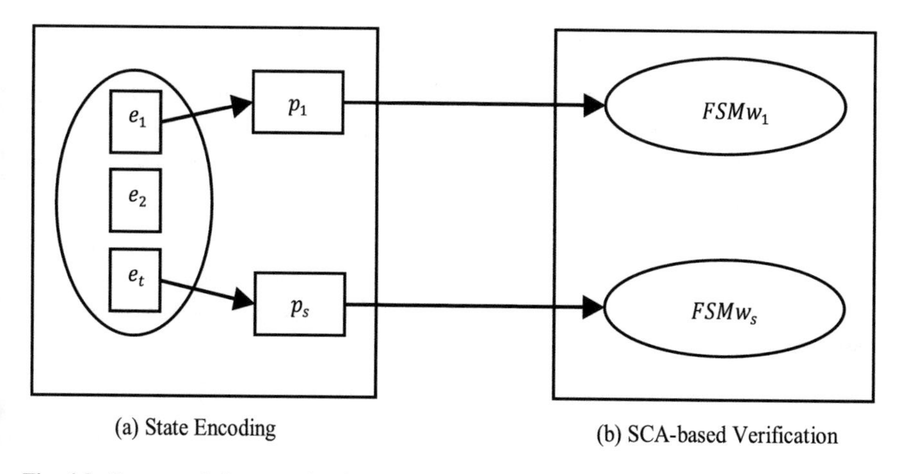

Fig. 4.8 Conceptual diagram of embedding and verification

"$n = m$" and this is known as "hot encoding," where each state is implemented by a distinct flip-flop. Another possibility is to use a "minimum" number of flip-flops, which will guarantee the usage of minimum storage elements ($m = \lceil \log_2 n \rceil$). In general, $t = |\Omega| \neq s$, indicating that there is not always possible to produce a sequential property p_i by *every* state encoding e_i. This condition is indicated by a pair of (e_t, p_s) in Fig. 4.8a.

We are interested in exploring state encoding solutions for nonredundancy. In a broad sense, this attempt requires not only avoiding any modification of the original FSM, but also using the minimum number of flip-flops. So, we will use the minimum number of flip-flops in this work. *However, the basic idea of the proposed solution is general and it can be extended for any value of "m", including " $m \geq n$".*

Example 4.5 (n = 6, m = 3): For M in Fig. 4.5, using the minimum number of flip-flops, $|\Omega| = t = 8 \times 7 \times 6 \times 5 \times 4 \times 3 = 20, 160$, there are six states (A, B, C, D, E, F), three flip-flops, and eight binary codes {000, 001, 010, 011, 100, 101, 110, 111} are available for the state encodings.

Hierarchical State Encoding: In a hierarchical (h = degree of hierarchy) encoding, multiple state encoding steps are applied. A two-tier state encoding (h = 2) is considered in this work, since this is consistent with the two-tier architecture (i.e., Fig. 4.4) The basic idea can be expanded to a multitier architecture, if needed.

$$e_i = e_i^1 + e_i^2 + \cdots + e_i^h = \bigcup_{j=1}^{h} e_i^j$$

In a two-tier state encoding, $e_i = e_i^1 + e_i^2$. In the initial state encoding (e_i^1), the goal is to extract a sequential property. Generally, in a tier-1 state encoding, a set of *blocks* of state is encoded. During the next step of state encoding (e_i^2), the number of *states* in *each block* is encoded. The final state encoding e_i is then the *concatenation* of e_i^1 and e_i^2.

Example 4.6 [Two-tier State Encoding]: For M in Fig. 4.5, $e_i^1 =$ {(α, 00), (β, 01), (γ, 10)} and $e_i^2 =$ {(a, 0), (b, 1)}. Then, $e_i =$ {(A, 000), (B, 001), (C, 010), (D, 011), (E, 100), (F,101)}.

Watermarked FSM: A watermarked FSM (FSM_w) can be constructed as a *state machine* using the tier-1 state encoding e_i^1. In general, there is a one-to-one correspondence between e_i^1 and FSM_w, meaning that, for a given e_i^1, we can construct *the* corresponding state machine. The detailed process is provided in Sect. 3.2.

Example 4.7 [Watermarked FSM]: For M in Fig. 4.5, $S_w = \{\alpha, \beta, \gamma\}, I_w = \varnothing, \delta_w(\alpha, a) = \beta, \delta_w(\beta, a) = \gamma, \delta_w(\gamma, a) = \alpha \, for \, a \in I$.
$e_i^1 : \alpha \rightarrow 00; \beta \rightarrow 01; \gamma \rightarrow 01$. Thus, this defines a state machine which becomes the watermarked FSM FSM_w..

4.3.2 Watermarked FSM

The process for embedding a watermarked FSM is described in Fig. 4.9. The input to the algorithm is an "n"-state FSM. The output is a watermarked FSM, FSM_w, which possesses the property P*. Note that FSM_w is a *state machine* with a triplet $FSM_w = (S_w, I_w, \delta_w)$. The most critical step (*Step 1*) is to find the maximal input-independent periodicity p_{max}.

The algorithm of finding p_{max} is shown in Fig. 4.10. The underlying idea of finding p_{max} is based on identifying the nontrivial input-independent closed partition with maximal cycle [19]. The complexity of the algorithm for finding p_{max} is $O(n^2)$, a polynomial time, since a pairwise state operation is required for the n states $(i.e, |S| = n)$ in *Step 1*. *Step 3* shows that the number of blocks in the smallest closed partition π_{min} is q ($< n$) and the nontrivial input-consistent partition τ can be set to π_{min} since this equality satisfies the condition $\pi_i \geq \tau$ in *Step 1*. Both determining the smallest closed partition in *Step 2* and finding the number of blocks in π_{min} in *Step 4* are straightforward and can be done in $O(n)$. Overall, the algorithm for extracting the watermarked FSM is efficient (i.e., at least a polynomial time).

Input: a "n"-state reduced FSM = (S, I, O, δ, λ),$|S| = n$
Requirement: $m = \lceil \log_2 n \rceil$ /* minimum no. of binary codes */
Output: a watermark $FSM_w = (S_w, I_w, \delta_w)$ /* state machine */

Procedure:
1. Find the *maximal* input-consistent periodicity p_{max}
2. Construct a state machine $FSM_w = (S_w, I_w, \delta_w)$ as follow:
3. The set of states: $S_w = \{s_1^*, s_2^*, \ldots, s_{p_{max}}^*\}$
4. The input: $I_w = \Phi$
5. The state transition: $\delta_w(s_1^*, a) = s_2^*$, $\delta_w(s_2^*, a) = s_3^*$, $\delta_w(s_3^*, a) = s_4^*$,....., $\delta_w(s_{p_{max}}^*, a) = s_1^*$ for $\forall a \in I$

Fig. 4.9 Pseudocode for extracting a watermarked FSM (FSM_w)

Input: a "n"-state reduced FSM = (S, I, O, δ, λ),$|S| = n$
Output: p_{max} /* Maximal periodicity */

Procedure:
1. Find a set of a closed partition $\pi = \{\pi_1, \pi_2, \ldots\ldots \pi_\varphi\}$ and a *nontrivial* input-consistent partition τ on S, where $\pi_i \geq \tau$, for $i = 1, 2, \ldots\ldots\phi$.
2. Determine the *smallest* closed partition π_{min} from $\pi = \{\pi_1, \pi_2, \ldots\ldots \pi_\varphi\}$.
3. Let $\pi_{min} = \{B_1; B_2; \ldots\ldots; B_q\} = \tau$.
4. The number of blocks in π_{min}, q, is the *maximal* periodicity p_{max}.

Fig. 4.10 Pseudocode for finding the maximal periodicity (p_{max})

Example 4.8 [Maximal Periodicity]: For M in Fig. 4.5, the maximal periodicity is $p_{max} = 3$ since $\pi_{min} = \tau = \{\overline{A,B}; \overline{C,D}; \overline{E,F}\} = \{\alpha;\beta;\gamma\} = \pi_1$ and the number of blocks in π_{min} is 3. Note that there exists other candidates of input-consistent partitions, namely, $\pi_2 = \{\overline{A,B,C,D}; \overline{E,F}\} = \{\overline{\alpha,\beta}; \overline{\gamma}\}$, $\pi_3 = \{\overline{A,B,E,F}; \overline{C,D}\} = \{\overline{\alpha,\gamma}; \overline{\beta}\}$, $\pi_4 = \{\overline{C,D,E,F}; \overline{A,B}\} = \{\overline{\beta,\gamma}; \overline{\alpha}\}$, and $\pi_5 = \{\overline{A,B,C,D,E,F}\} = \{\overline{\alpha,\beta,\gamma}\}$. However, π_5 is a *trivial* input-consistent partition since it combines all states into a single block, which implies that no useful information is processed with this state partition. All other partitions above are nontrivial input-consistent partitions. However, the *smallest* closed partition is π_1.

Hierarchical (Two-tier) State Encoding: Upon determining the partition π_{min} and the maximal periodicity p_{max}, a state encoding e_i can be performed by assigning the minimum number of binary codes to the blocks of the smallest partition. In a tier-1 state encoding, the assignment with a minimum number of binary bits can be made on the blocks of state $\{B_1; B_2; \ldots\ldots; B_q\}$ in π_{min}. This is described in *Steps 1* and *2* in Fig. 4.11. Pseudocode of tier-1 state encoding (*Steps 1* and 2) for the blocks in the smallest closed partition is associated with FSM_w. Pseudocode of tier-2 state encoding (*Steps 3, 4* and 5) for the blocks of partition μ is associated with the residual FSM_r. Once the smallest closed partition (π_{min}) and nontrivial input-consistent partition (τ) are determined, it is simple to perform the initial state encoding (e_i^1) using the minimum number of binary bits (*Steps 1* and 2). The complexity of tier-1 state encoding procedure is $O(q)$. Note that *Step 2* might not produce a unique state encoding.

Example 4.9 [State Encoding; Tier-1]: From Example 4.8, $q = 3, m_w = 2, e_i^1 = \{(\alpha, 00), (\beta, 01), (\gamma, 10)\}$. Note that there exists other tier-1 state encodings possible such as $\{(\alpha, 01), (\beta, 10), (\gamma, 11)\}$.

The tier-1 state encoding e_i^1 above defines a watermarked FSM (FSM_w) which is a three-state FSM performing the computation to distinguish three blocks of states in the partition of $\{\overline{A,B}; \overline{C,D}; \overline{E,F}\} = \{\alpha;\beta;\gamma\}$. To complete the state encoding e_i, a tier-2 state encoding e_i^2 needs to be done. Informally, e_i^2 should distinguish each state in every block B_i *in*$\{B_1; B_2; \ldots\ldots; B_q\}$. This can be done with another partition $\mu = \{\overline{A,C,E}; \overline{B,D,F}\}$, as an example.

A general procedure for a tier-2 state encoding e_i^2 is described in Steps (3)–(5) in Fig. 4.11. Given $\pi_{min,}$ finding another partition μ satisfying the condition (i.e.,

Input: State partition $\pi_{min} = \tau = \{B_1;\ B_2; \ldots\ldots;\ B_q\}$, $q < n$.
Requirement: Use the minimum number of binary codes
Output: State encoding of (e_i^1) and (e_i^2)

Procedure:

1. Let $m_w = \lceil \log_2 q \rceil$
2. Perform one-to-one mapping using m_w *binary bits* on $\{B_1;\ B_2; \ldots\ldots;\ B_q\}$.
3. Find another state partition μ such that $\pi_{min} \cdot \mu = \emptyset$; Let $\mu = \{b_1;\ b_2; \ldots.;\ b_r\}$.
4. Let $m_r = \lceil \log_2 r \rceil$
5. Perform one-to-one mapping using m_r *binary bits* on $\{b_1;\ b_2; \ldots.;\ b_r\}$.

Fig. 4.11 Pseudocode of tier-1 and tier-2 state encoding

$\pi_{min} \cdot \mu = \varnothing$) is straightforward since it can split the states in each block of π_{min} into an individual state. *Step 3* can be done in polynomial time, $O(q \cdot |B_i|)$, where q is the number of blocks in π_{min} and $|B_i|$ is the maximum number of states in the block. Steps 4 and 5 are similar to the tier-1 state encoding procedure. Example 4.10 shows the results of tier-2 and final state encodings. The final state encoding e_i is made using the *concatenation* of two tiers of state encodings e_i^1 and e_i^2.

Example 4.10 From Examples 4.8 and 4.9, $\mu = \{\overline{A, C, E}; \overline{B, D, F}\} = \{b_1; b_2\}$ since $\pi_{min} \cdot \mu = \varnothing$. Also, $r = 2, m_r = 1, e_i^2 = \{(b_1, 0), (b_2, 1)\}$. For the final state encoding:

$$e_i = e_i^1 + e_i^2 = \{(A, 000), (B, 001), (C, 010), (D, 011), (E, 100), (F, 101)\}.$$

4.4 Analysis

In this section, we provide the analysis of the proposed approach. The limitation of the proposed approach and the potential solution are also discussed.

4.4.1 Existence of Watermarked FSM

We provide the analysis for the existence of the watermarked FSM possessing the sequential property P* [19].

Theorem 4.1 *The existence of a closed partition π and a nontrivial input-consistent partition τ_i on S in a reduced FSM = (I, O, S, δ, λ), where $\pi \geq \tau_i$, is a necessary and sufficient condition for the existence of the watermarked $FSM_w = (S_w, I_w, \delta_w)$ possessing the property P*.*

Proof (Necessary part) If the FSM_w possesses a closed partition π such that $\pi \geq \tau_i$, then, for a given state S_i and every input in *I*, the next states must be in the same block of τ_i, and therefore in the same block of π. Consequently, for a given initial state, the block of π in which the state of FSM_w is contained after any finite input sequence depends only on the initial block and on the length of the sequence.

(Sufficient part) If such a watermarked FSM possessing P* exists, there must exist a cycle of states such that $\delta_w(s_1, x) = \delta_w(s_2, x) = \ldots = \delta_w(s_p, x)$ for any input x in $I_w = I$. Then, $\{s_1; s_2; \ldots; s_p\} = \pi = \tau_i$, and the maximal periodicity ($p_{max} = p$) of a cycle can be chosen.

4.4.2 Analysis of Proposed Approach (Qualitative)

Based on the comprehensive set of requirements for desirable IP watermarks described in [11], we analyze the proposed approach qualitatively. As a simple measure, the compliance of each requirement is indicated by a scale of "H (high compliance)", "M (middle)", or "L (low)".

- ***Not relying on the secrecy of the algorithm***: one of the oldest security principles defined by Kerckhoffs [20] says "The system must not require secrecy and can be stolen by the enemy without causing trouble." To protect the authorship, the algorithm should depend on the system properties instead of the secrecy of the algorithms. The proposed approach is based on extracting system's sequential properties (P*) that exist inherently in the system [H].
- ***Level of reliability***: this is further divided into two requirements of robustness and false positives. Robustness measures the strength of the hidden watermark against attacks. False positives occur when the detector can find the watermark in a nonwatermarked design. Both of them are addressed in the following section (Attack Analysis, Sect. 4.3). The proposed approach satisfies both requirements [H].
- ***Affecting the design functionality***: watermarking techniques should prove their soundness against affecting the behavior of original system. In the proposed approach, the FSM is decomposed into two component FSMs, FSM_w and FSM_r, where the original functionality is always preserved [H].
- ***Preventing an intruder from re-embedding another watermark***: this is to provide the authenticity of the watermark. Watermark designers need to find techniques to protect their designs from intruders to embed another watermark in the design. In the proposed scheme, it is likely that the functionality of an FSM may change and thus detectable, if another watermark is added. However, the intruder may try to identify another inherent property to create new watermark. Thus, this requirement may or may not be satisfied [Not conclusive].
- ***Embedding enough data to identify ownership***: the watermarking scheme should add enough data to identify the owner of the design. In the proposed scheme, the maximality (p_{max}) of periods was used to embed more data. However, the amount of data at an FSM level (not a state level) can be relatively small. Note that this requirement is related to another requirement of "Implementation overhead" below [L].
- ***Implementation overhead:*** compared to the original design without watermarking, the watermarking scheme should not introduce (or minimize) overheads, especially in the area, power, and delay overheads. The proposed approach does not require additions of neither new states/transitions nor new verification circuitry [H].
- ***Detection and tracking:*** watermark embedding is only half the process, detection (i.e., verification) is the second important aspect in any watermarking technique. The feasibility of SCA-based detection mechanism was demonstrated for a set of small-size FSMs [21, 22] [M].

- ***Asymmetry:*** Sharing IP designs poses the same threats as other secret data in the public domain [11]. Third parties such as brokers and subcontractors need to know the watermark key for tracking purposes. But, these parties are not considered trustworthy entities. It was demonstrated that asymmetric watermarking with the watermark key *can* be supported [22] [H].

Despite the existence of some deficiency, the proposed approach satisfies most of requirements at a reasonable level. It would not be technically feasible, if not impossible, to build a single watermarking solution which can satisfy *all* of the requirements at once.

4.4.3 Attack Analysis

IP watermarking attacks can be further categorized in the main classes below, based on the work [23].

- ***Removal Attacks***: Removal attacks are divided into either elimination attacks or masking attacks. In elimination attacks, the watermark can be eliminated completely by an attacker. For instance, the attacker tries to estimate the watermark and subtract it from the watermarked design. In the proposed approach, it will be technically infeasible to eliminate the watermarked FSM without affecting the original functionality since the watermarked FSM itself performs the subcomputation. On the other hand, masking attacks aim at distorting the watermark detector to disable its ability to sense the presence of the watermark. This is a verification related. It would be very difficult to disable the SCA-based verification process, since it is based on contactless verification [21, 22].
- ***False Positive*** (Probability of Coincidence or Watermark Collision): the probability of coincidence was defined as "the odds that an unintended watermark is detected in a design" [14]. Often, this is used as a measure of watermark validity. It was demonstrated using a set of small FSMs that the collisions practically may not be a problem [21, 22].
- ***Embedding Attacks (Forging)***: Similar to "Preventing intruders from re-embedding another watermark" (Sect. 4.2).

4.4.4 The Limitation and Solutions

The proposed approach is based on extracting the inherent sequential property $P*$ by a hieratical state encoding method. However, if such a property cannot be extracted due to the inexistence of the property in a given FSM, the proposed approach cannot be used. The existence of $P*$ is based on specific conditions. The necessary and sufficient conditions are provided in Theorem 4.1.

If a given FSM cannot be decomposed into a watermarked FSM possessing P*, we should consider using a weaker condition (e.g., relaxing *maximality*). But, this solution will be less desirable since it will further reduce the embedded data (see 4.2 —*"Embedding enough data to identify ownership")*.

More practical solution would be to remove the condition of input independence. By doing so, the IP designer can use the specific input values as part of IP designer's secret (e.g., watermark input signature). In this case, new or refined embedding algorithms will need to be developed.

4.4.5 Discussions

The readers might be interested in using a specific state encoding as part of the designer's IP: the encoding space $\Omega = \{e_1, e_2, \ldots, e_t\}$ is quite huge and randomly guessing a particular encoding can be a difficult task. However, we did not choose this for the following reasons: first, despite the colossal size of the state encoding space ($|\Omega| = very\, big$), collisions may occur (see *State Encoding Tier-1;* Example 4.9). Second, we believe that using the system's inherent properties can be more effective in the watermarking application. Another interesting discussion is to see if more effective sequential properties (instead of a maximal cyclic behavior) can be used.

4.5 Conclusions

Historically, the problem of state encoding (or state assignment) in sequential machines, or more generally a regular sequential function, was extensively studied in 1960s. However, the dominant application areas of the state encoding problem were sequential circuit optimizations in later decades (i.e., especially during 1980s and 1990s.) We believe that it can be useful to apply these well-studied areas of state encoding schemes to other or new application areas, such as FSM IP watermarking, especially as a way of providing IP protection in embedded controller designs. This attempt can be beneficial if the IP protection solution can be achieved without adding any functional overheads (e.g., adding dummy states.).

In this chapter, we described one of the possible approaches for protecting the design IPs of an embedded controller. For a class of FSMs, based on cascade machine decomposition and the hierarchical state encoding scheme, we presented the different types of watermark embedding method which does not require adding dummy states and/or state transitions. We analyzed the proposed method, which include addressing both the strengths and the weaknesses, based on the criteria used in the field of hardware IP protection. Despite the existence of some limitation, we found that the proposed schemes are favorably evaluated in most of the criteria used in the field of IP watermarking.

Acknowledgement The work presented in this paper was realized in the frame of the SALWARE project number ANR-13-JS03-0003 supported by the French" Agence Nationale de la Recherche" and by the French "Fondation de Recherche pour l'Aéronautique et l'Espace".

References

1. M. Tehranipoor, C. Wang (eds.), *Introduction to Hardware Security and Trust* (Springer Science+Business Media, LLC, 2012), Chaps. 6 and 17
2. L. Bossuet, D. Hely, Salware: salutary hardware to design trusted IC, in *Proceedings of the Trustworthy Manufacturing and Utilization of Secure Devices Workshop* (TRUDEVICE '13), Avignon, France (2013), pp. 30–31. http://hal-ujm.ccsd.cnrs.fr/ujm-00833871
3. G. Wolfe, J.L. Wong, M. Potkonjak, Watermarking graph partitioning solutions. IEEE Trans. Comp.-Aided Des. Integ. Cir. Sys. **21**, 1196–1204 (2002). doi:10.1109/TCAD.2002.802277
4. J.L. Wong, G. Qu, M. Potkonjak, Optimization-intensive watermarking techniques for decision problems. IEEE Trans. Comp.-Aided Des. Integ. Cir. Syst. **23**(1), 119–127 (2006). doi:10.1109/TCAD.2003.819900
5. G. Qu, M. Potkonjak, Analysis of watermarking techniques for graph coloring problem, in *Proceedings of the IEEE/ACM international conference on Computer-aided design* (ICCAD '98) (ACM, New York, NY, USA, 1998), pp. 190–193. doi:10.1145/288548.288607
6. L. Yuan, G. Qu, Information hiding in finite state machine, in J. Fridrich (ed.), *Proceedings of the 6th international conference on Information Hiding* (IH'04) (Springer-Verlag, Berlin, Heidelberg, 2004), pp. 340–354. doi:10.1007/978-3-540-30114-1_24
7. J. Lach, W.H. Mangione-Smith, M. Potkonjak, Robust FPGA intellectual property protection through multiple small watermarks, in M.J. Irwin (ed.) *Proceedings of the 36th annual ACM/IEEE Design Automation Conference* (DAC '99) (ACM, New York, NY, USA, 1999), pp. 831–836. doi:10.1145/309847.310080
8. F. Koushanfar, Y. Alkabani, Provably secure obfuscation of diverse watermarks for sequential circuits, in *Proceedings of the International Symposium on Hardware-Oriented Security and Trust* (HOST '10), (2010), pp. 42–47. doi:10.1109/HST.2010.5513115
9. B. Le Gal, L. Bossuet, Automatic low-cost IP watermarking techniques based on output mark insertions, in *Design Automation for Embedded Systems*, vol. 16, issue 2 (Springer Science +Business Media, 2012), pp. 71–92. doi:10.1007/s10617-012-9085-y
10. F.A. Petitcolas, R.J. Anderson, M.G. Kuhn, Information hiding—a survey, in *Proceedings of the IEEE*, 87(7), 1062–1078 (1999). doi:10.1109/5.771065
11. A.T. Abdel-Hamid, S. Tahar, E.M. Aboulhamid, A survey on IP watermarking techniques, in *Design Automation for Embedded Systems*, vol. 9 (Springer Science+Business Media, Berlin, 2004), pp. 211–227. doi:10.1007/s10617-005-1395-x
12. A.L. Oliveira, Robust techniques for watermarking sequential circuit designs, in Irwin, M. J. (ed.) *Proceedings of the 36th annual ACM/IEEE Design Automation Conference* (DAC '99) (ACM, New York, NY, USA, 1999), pp. 837–842. doi:10.1145/309847.310082
13. A.L. Oliveira, Techniques for the creation of digital watermarks in sequential circuit design. IEEE Trans. Comp.-Aided Des. Integ. Cir. Sys. 20(9), 1101–1117 (2001). doi:10.1109/43.945306
14. I. Torunoglu, E. Charbon, Watermarking-based copyright protection of sequential functions. IEEE J. Solid-State Circuits **35**(3), 434–440 (2000). doi:10.1109/4.826826
15. F. Koushanfar, I. Hong, M. Potkonjak, Behavioral synthesis techniques for intellectual property protection. ACM Trans. Des. Autom. Electron. Syst. 10, 3, 523–545 (2005). doi:10.1145/1080334.1080338
16. A. Cui, C.H. Chang, S. Tahar, A.T. Abdel-Hamid, A robust FSM watermarking scheme for IP protection of sequential circuit design. IEEE Trans. Comp.-Aided Des. Integ. Cir. Sys. 30(5), 678–690 (2011). doi:10.1109/TCAD.2010.2098131

17. M. Lewandowski, R. Meana, M. Morrison, S. Katkoori, A novel method for watermarking sequential circuits, in *Proceedings of the IEEE International Symposium on Hardware-Oriented Security and Trust* (HOST'12), San Francisco, CA, pp. 21–24 (2012). doi:10.1109/HST.2012.6224313
18. J. Hartmanis, R.E. Stearns, *Algebraic Structure Theory of Sequential Machines* (Prentice-Hall International Series in Applied Mathematics) (Prentice-Hall, Inc. (1996), Upper Saddle River, NJ, USA
19. Z. Kohavi, *Switching and Finite Automata Theory*, 2nd edn. (McGraw-Hill, 1978)
20. A. Kerckhoffs, La Crytographie Militaire. Journal des sciences militaires, 9, 5–38 (1883). 161 –191 (February 1883)
21. E. Jung, S. Choi, Identification of IP control units by state encoding, in IEEE Computer Society Annual Symposium on VLSI, July 2015, pp. 216–220. doi:10.1109/ISVLSI.2015.43
22. C. Marchand, L. Bossuet, E. Jung, IP watermarking verification based on power consumption analysis, in *Proceedings of the 27th IEEE International System-on-Chip Conference (SOCC '14), Las Vegas*, Sept 2014, pp. 330–335. doi:10.1109/SOCC.2014.6948949
23. I.J. Cox, M.L. Miller, J.A. Bloom, C. Honsinger, *Digital watermarking* (Morgan Kaufmann Publishers, 1998)

Chapter 5
Side Channel Analysis, an Efficient Ally for IP Protection

Lilian Bossuet and Cédric Marchand

5.1 Introduction

One of the solutions for the IP designers to protect their intellectual property is to be able to detect the presence of a copy of an IP embedded in a digital device by using IP identification. Works on IP watermarking and IP fingerprinting try to provide the IP identification service [1]. But, most of the time the published solutions are not practical mainly because of the complexity of the watermarking/fingerprinting verification scheme [2, 3]. If watermarking and fingerprinting are efficient to prove an illegal copy of IP, the industrial detection of integrated circuit counterfeiting in the supply chain needs rapid and contactless checking of intellectual property data. For both these applications, providing a secure and discreet mean of data communication is a challenging work. We think that the use of side channel analysis, usually performed for physical attacks on cryptographic algorithms implementation, could be an efficient solution. The power consumption and the electromagnetic radiation of an integrated circuit are two interesting side channels for IP watermarking verification and discreet transmission of intellectual property data. This chapter presents two works that try to exploit these two side channels in the case of IP protection.

5.1.1 *Side Channel Analysis*

In the area of security, the techniques used to attack and to defend have always been similar and the means designed for attacks can sometimes be used for protection. A well-known threat in cryptographic engineering is side channel analysis

L. Bossuet (✉) · C. Marchand
Laboratoire Hubert Curien, Jean Monnet University, Saint-Etienne, France
e-mail: lilian.bossuet@univ-st-etienne.fr

L. Bossuet and L. Torres (eds.), *Foundations of Hardware IP Protection*,
DOI 10.1007/978-3-319-50380-6_5

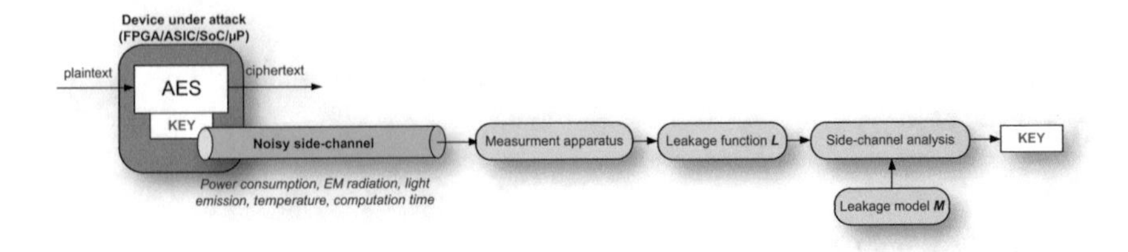

Fig. 5.1 General presentation of the side channel analysis of the hardware implementation of a cipher (or decipher)

(SCA) [4, 5]. SCA are widely used in cryptographic engineering as passive attacks because they make it possible to retrieve secret information (such as secret keys) with relatively few measurements and sometimes using inexpensive equipment. SCA attacks work even when the algorithm has been shown to be robust against algebraic cryptanalysis. Most of the dynamic characteristics of both hardware and software implementations of cryptographic primitives can be used for side channel analysis: computation time, power consumption, electromagnetic radiation, optical radiation, even the sound produced during computation. These physical quantities are thus widely exploited during SCA that aim to understand the circuits behavior (or to discover the secret information they contain, such as the secret keys required by the encryption/decryption process) by jointly analyzing the cipher algorithm, the data it produced (cipher texts) or received (plaintexts) and the information leaked by the side channel. This analysis is typically performed using statistical tools. The Fig. 5.1 gives us the general presentation of a SCA targeting a hardware implementation of a cipher (the symmetric cipher AES in the Fig. 5.1).

However, the techniques used for side channel analysis can also be used to implement a salutary hardware [6] for IP protection: e.g., for reading intellectual property data from the device or for device authentication (watermark checking).

5.1.2 Use Cases for IP Protection

Reading hidden data of intellectual property (such as watermarking and/or fingerprinting) can be used in two ways during the lifetime of an integrated circuit. First, the system integrator or the end user may want to check if all the items in a set of integrated circuits he/she purchased are authentic. To do so, they check a watermark on all integrated circuits, and authenticate each one of them as illustrated in the top half of Fig. 5.2. Alternatively, a designer can check that his IP has not been copied and illegally integrated in another system. To do so, he/she can compare a watermark he/she owns and the one on the integrated circuit, as illustrated in the bottom half of Fig. 5.1. If they are identical, he/she can prove that the integrated circuit manufacturer has used his IP illegally.

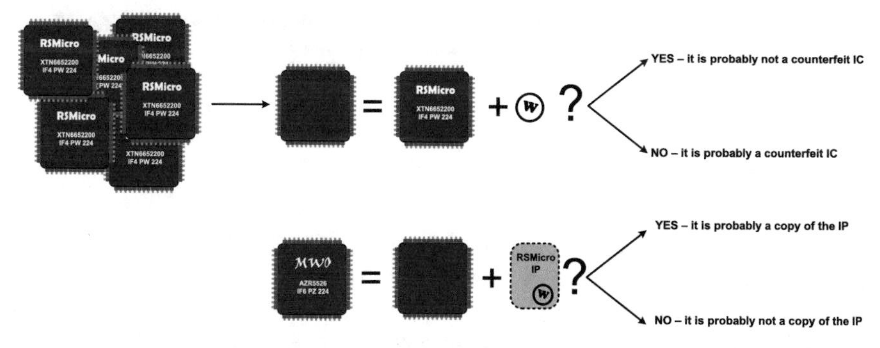

Fig. 5.2 Two scenarios of IP protection, (*top*) detection of integrated circuit counterfeiting from a set of identical referenced integrated circuits and (*bottom*) cloning detection of an integrated circuit or an IP from a device of a competitor

Some published works propose spy circuitry using side channels to identify the embedded intellectual property (hidden transmission of IP information, such as watermark or PUF response, on the side channel). For example in [7], the thermal channel representing a contactless communication was used to transfer information from an embedded tag to a remote receiver. However, the embedded thermal tag used in this commercial solution requires a relatively large area (255 Spartan-3 slices). In [8], the authors propose to use two shift registers to generate a recognizable signature-dependent pattern in the power consumption to reveal the IP signature. Power consumption was also used in [9] to communicate the IP watermark data using classical differential power analysis (DPA [4]). To reinforce such work, the authors of [10] propose to use the power supply signal of an IP as a physical hash function for fingerprinting.

This chapter completes this rapid state of the art by presenting a scheme of FSM watermarking verification using power consumption analysis (Sect. 5.2) and the high-speed and contactless transmission of IP information by using the electromagnetic channel (Sect. 5.3).

5.2 FSM Watermarking Verification Scheme Using Power Consumption Analysis

In this section, we address the problem of verification of watermarked finite state machines (FSMs) presented in the Chap. 4 of this book. The verification scheme of the IP watermark uses a correlation analysis based on the measurement of the power consumption of an IC. In order to make this verification possible, a lightweight component which amplifies the side channel leakage is added to the IP. This component only highlight the state transition of the FSM by bringing nonlinearity but does not interfere with the working FSM. In addition, it reduces the risk of collision between different IPs with the same FSM. A preliminary version of this work was presented during the conference SOCC 2014 [3].

5.2.1 Principle

The purpose of this section is to demonstrate that it is possible to verify a watermarked finite state machine (FSM) using SCA. To perform the analysis, the owner of the IP provides a device produced in a trusted manner (i.e., the device contains only what the owner created without alterations). This device is called reference device (*RefD*) and is used as reference in the watermarked FSM verification process. Then, the goal is to determine if a device under test (*DUT*) contains the same watermarked FSM as the *RefD*. The power consumption of this *RefD* is compared to the power consumption of the *DUT* using the proposed correlation methods.

As presented in Chap. 4 of this book, the watermarked FSM is input-consistent (i.e., input-independent) and cyclic, so it is not necessary to verify state transitions in the two devices using a specific input sequence. Indeed, if the state sequence resulting from one input contains more transitions than the watermarked FSM periodicity, then verification is possible. The same input sequence is sent to the two devices to be sure that the state sequence has the same length for the *RefD* and for the *DUT*.

To verify if a *DUT* contains the same watermarked FSM as the *RefD*, a large number n of power consumption traces are measured on the *RefD* and grouped in a set called T_{ref}. The same number n of power traces are measured on the *DUT* and grouped in the set called T_{DUT}. Then k traces of T_{ref} are selected using a function which randomly selects k distinct elements uniformly inside a set X (noted $U_X(k)$). This function can be defined as follows:

$$\begin{aligned} &\forall k \in [\,[1;\ n]\,], U_X(k) = \{e_1, e_2, \ldots, e_k\}, \\ &\text{such that } \forall i,\ j \in [\,[1,\ k]\,],\ i \neq j \Leftrightarrow e_i \neq e_j \end{aligned} \tag{5.1}$$

The mean of selected traces is calculated and used as a unique reference to the correlation computation. This averaged trace is noted A_{ref} and defined as $A_{ref} = mean(U_{T_{ref}}(k))$. The same operation is repeated to calculate a number m of k-averaged traces with the set T_{DUT}. The set noted A_{DUT} contains these m k-averaged traces and is defined by

$$A_{DUT} = \{mean(U_{T_{DUT}}(k))\}_m. \tag{5.2}$$

When all k-averaged traces are calculated, the correlation between A_{ref} and each element of A_{DUT} is computed using the Pearson coefficient defined by:

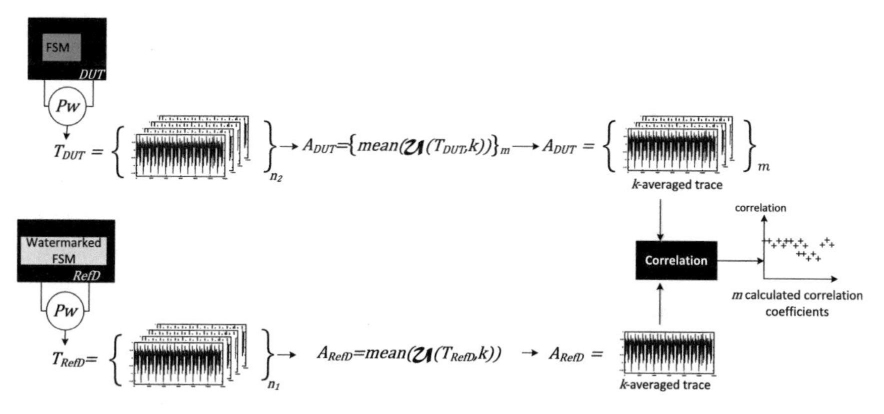

Fig. 5.3 Correlation calculation flow with one device under test (*DUT*) and one reference device (*RefD*)

$$\rho(x,y)=\frac{\sum_{i=1}^{l}(x_i-\bar{x})\cdot(y_i-\bar{y})}{\sqrt{\sum_{i=1}^{l}(x_i-\bar{x})^2\cdot\sum_{i=1}^{l}(y_i-\bar{y})^2}}, \tag{5.3}$$

where x and y are two traces of length l and $\bar{x}$ is the mean of x. Since only one k-averaged trace for the *RefD* and m k-averaged traces for the *DUT* are used; all variations of the computed correlation coefficients must be due to the *DUT* and not to the *RefD*.

The result of this process is a set of m correlation coefficients. Figure 5.3 shows this calculation process with a schematic. When all m coefficients are computed, the analysis is performed using statistical tools such as the mean or the variance of the correlation coefficients.

5.2.2 Experimental Results

By using the correlation computation process previously described, some experiments are performed in order to verify a watermarked FSM in a real device. For the following experiment, four FSMs are designed. The first one, called *FSM_A*, is an 8-bit binary-counter. The second FSM (*FSM_B*) is an 8-bit Gray-counter. An AES *Sbox* is added to the *FSM_A* and *FSM_B* to create *FSM_C* and *FSM_D*, respectively. *Sbox* uses an 8-bit input. An additional watermark is fixed to the same randomly chosen value W_k in *FSM_C* and *FSM_D*. Figure 5.4 shows these four reference FSMs with schematics.

The four FSMs are implemented inside four Altera Cyclone III FPGAs in order to create four different *RefDs* (*FSM_A*, *FSM_B*, *FSM_C*, *FSM_D*). Four *DUTs* ($DUT_{\#1}$, $DUT_{\#2}$, $DUT_{\#3}$, $DUT_{\#4}$) are created by implementing the same FSMs in

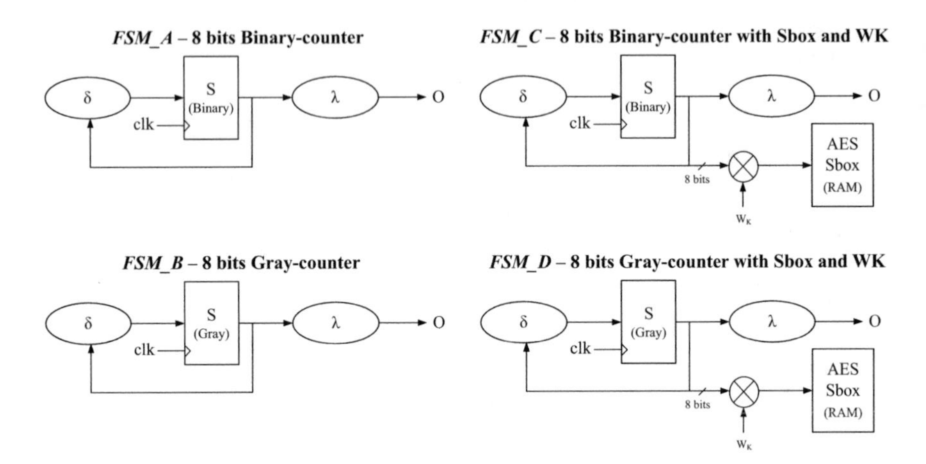

Fig. 5.4 The four tested FSMs

four other Altera Cyclone III FPGAs. Note that these experimental results are also obtained using only one Cyclone III for all measurements. Therefore, the proposed work is insensitive to the CMOS variation process [3].

For each *RefDs*, 10,000 power consumption traces are measured to create the following sets of traces: T_{FSM_A}, T_{FSM_B}, T_{FSM_C} and T_{FSM_D}. The same measurements are performed with the four *DUTs* to create the following sets: $T_{DUT_{\#1}}$, $T_{DUT_{\#2}}$, $T_{DUT_{\#3}}$ and $T_{DUT_{\#4}}$.

For this experiment, correlation coefficients are calculated with $k = 50$ and $m = 20$ (see [3] for the selection methodology of these parameters). Using a set of power consumption traces T_{FSM_X} with $X \in \{A, B, C, D\}$, the averaged reference trace A_{FSM_X} is defined by $A_{FSM\_X} = mean(U_{T_{FSM_X}}(50))$. And using a set of measured traces $T_{DUT_{\#Y}}$ $(Y \in \{1, 2, 3, 4\})$, the set $A_{DUT_{\#Y}}$, which contains 20 50-averaged traces, is defined by $A_{DUT_{\#Y}} = \left\{mean(U_{T_{DUT_{\#Y}}}(50))\right\}_{20}$.

Correlation coefficients are calculated between A_{FSM_X} and each trace of $A_{DUT_{\#Y}}$. The resulting set of 20 correlation coefficients is noted $C_{XY,50}$. For generic parameters k and m, the set $C_{XY,k}$ is defined by $C_{XY,k} = \{\rho(A_{FSM_X}, A_{DUT_{\#Y}}(i)),$ $i \in [.15em[1, m].15em]\}$.

Figure 5.5 shows these calculated sets of correlation coefficients grouped by *RefD*. For $X \in \{A, B, C, D\}$, the sub-picture titled *FSM_X* presents the following sets of correlation coefficients:$C_{X1,\ 50}$, $C_{X2,\ 50}$, $C_{X3,\ 50}$ and $C_{X4,\ 50}$. By observing this sub-picture, it is possible to determine intuitively which *DUT* contains the *FSM_X*.

In order to automatically determine which *DUT* contains which FSM, some distinguishers need to be considered. The means and the variance of the correlation coefficient sets are selected as potential distinguishers. Table 5.1 presents the mean of sets $C_{XY,\ k}$ (noted $\overline{C_{XY,\ k}}$) with $X \in \{A, B, C, D\}$ and $Y \in \{1, 2, 3, 4\}$ and Table 5.2 presents the variance of these sets (noted $var(C_{XY,\ k})$). In these two tables, the last

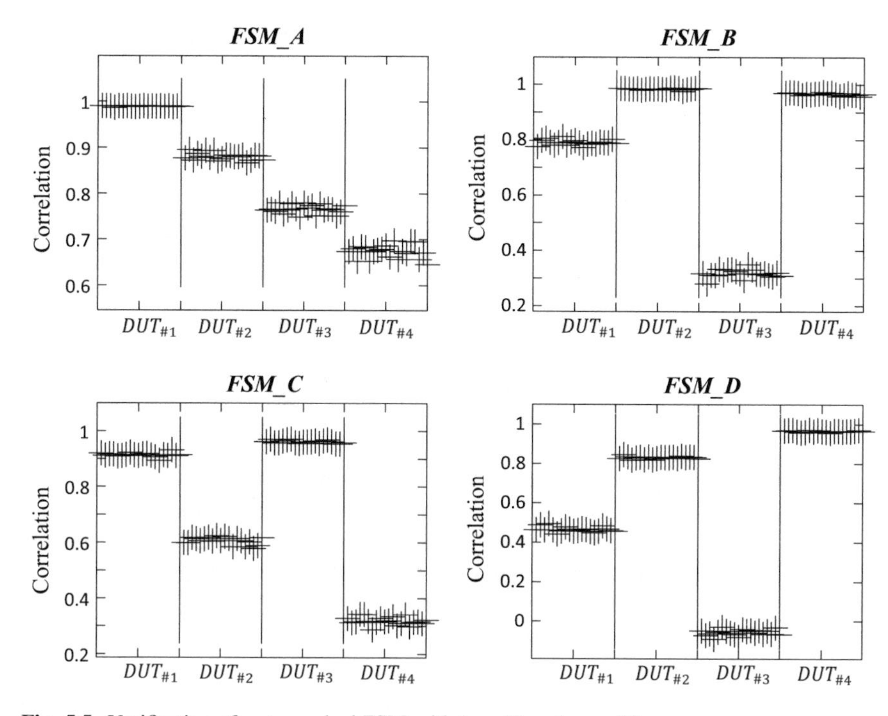

Fig. 5.5 Verification of watermarked FSM with k = 50 and m = 20

Table 5.1 Means of the different sets of correlation coefficients

	$DUT_{\#1}$	$DUT_{\#1}$	$DUT_{\#1}$	$DUT_{\#1}$	Δ_{mean} (%)
FSM_A	**0.989**	0.880	0.765	0.673	11
FSM_B	0.790	**0.983**	0.317	0.966	1.7
FSM_C	0.914	0.605	**0.962**	0.319	5
FSM_D	0.464	0.830	−0.057	**0.965**	14

Table 5.2 Variance of the different sets of correlation coefficients

	$DUT_{\#1}$	$DUT_{\#1}$	$DUT_{\#1}$	$DUT_{\#1}$	Δ_{var} (%)
FSM_A	**1.02e-6**	6.09e-5	8.86e-5	2.26e-4	98.3
FSM_B	9.22e-5	**8.40e-6**	2.26e-4	1.93e-5	56.5
FSM_C	5.28e-5	1.73e-4	**2.89e-5**	2.15e-4	36.4
FSM_D	1.68e-4	4.38e-5	2.34e-4	**1.83e-5**	58.2

column gives a confidence distance of the distinguisher for each row. This confidence distance is a percentage representing the distance of the distinguisher value between the two best *DUTs* according to one reference *FSM_X* ($X \in \{A, B, C, D\}$). In order to define this confidence distance, the followings functions are introduced:

- $max_2(E)$ is the function which returns the second highest element of the set E,
- $min_2(E)$ is the function which returns the second smallest element of the set E.

Considering the mean as distinguisher, the confidence distance (Δ_{mean}) is defined by:

$$\Delta_{mean}(\mathrm{X}) = 100* \left[1 - \frac{max_2(\{\overline{C_{XY,\,k}}; Y \in \{1,2,3,4\}\})}{max(\{\overline{C_{XY,\,k}}; Y \in \{1,2,3,4\}\})} \right] \quad (5.4)$$

For the variance, the confidence distance is noted Δ_{var} and defined by:

$$\Delta_{var}(\mathrm{X}) = 100* \left[1 - \frac{min(\{var(C_{XY,\,k}); Y \in \{1,2,3,4\}\})}{min_2(\{var(C_{XY,\,k}); Y \in \{1,2,3,4\}\})} \right] \quad (5.5)$$

For the two distinguishers, mean and variance, the higher the confidence distance, Δ_{mean} and Δ_{var} respectively, is the better the distinguisher is.

By using the variance as the distinguisher it is certain that $DUT_{\#1}$, $DUT_{\#2}$, $DUT_{\#3}$ and $DUT_{\#4}$ respectively contain *FSM_A*, *FSM_B*, *FSM_C* and *FSM_D*. Indeed, for each row the confidence distance is very high (from 36.4 to 98.3 % in Table 5.2). The use of the mean as a distinguisher is less efficient. The confidence distance is too small for *FSM_B* and *FSM_C* to be reliably distinguished (1.7 % and 5 % respectively in Table 5.1). Nevertheless, it is shown that taking the maximum value of the mean may distinguish the good *DUT* without error. But, in the presence of noise during measurement, incorrect results may be obtained by relying upon such a small confidence distance.

This experiment shows us that the variance is a more reliable distinguisher than the mean. This can be explained by the fact that all the studied circuits are strongly synchronized with the clock; thus, two devices with different FSMs inside can be highly correlated because of clock synchronicity. Using variance as a distinguisher is better because it highlights the transition between states better than the mean, which highlights the clock synchronicity.

5.3 Electromagnetic Communication of IP Data

Efficient IP identification scheme needs to be contactless, rapid and ultra-lightweight. Up to now, these three characteristics are not available in the state of the art. To meet these requirements, in this chapter we propose an ultra-lightweight binary frequency shift keying (BFSK) transmitter to forward IP identity (that could generated, for example by a feedback shift register or a physical unclonable function [11]) discreetly using an electromagnetic channel. Such circuit is usually called "spy circuitry." Using the electromagnetic channel, it is possible to contactless check the presence of an IP inside a digital device. A preliminary version of this work was presented during the conference VLSI-SOC 2015 [12].

5.3.1 Principle

Previous works on the electromagnetic attacks targeting true random number generators (TRNGs) showed that electromagnetic radiation can be used very efficiently for both active (fault injection [13]) and passive (side channel analysis [14]) attacks. Compared to power analysis, the attacker measuring the near-field electromagnetic emissions can obtain additional partial information about the device, since, unlike measurement of power consumption, electromagnetic radiation can be measured locally. One of the main advantages of this side channel is that it is impossible to hide the leak concerning electromagnetic radiation by using a global countermeasure. Moreover, the electromagnetic test bench is not expensive (less than USD 10K without an oscilloscope, which is the most expensive component). Last but not least, a spectral analysis of the electromagnetic radiation provides information on the oscillating structure, such as a ring oscillator [14]. For all these reasons, we use the electromagnetic channel for our IC/IP identification scheme. To this end, we designed an ultra-lightweight BFSK transmitter.

As mentioned above, salware and malware can be based on similar principles. The same is true for the proposed BFSK principle, which can be used to design both salware (i.e., IP identity transmitter) and malware (i.e., stolen data transmitter driven by a hardware Trojan), as illustrated in Fig. 5.6. There are two differences between using the BFSK as salware or malware. First, IP identification is activated outside the device by an ID checker, while the Trojan is activated internally. For example, the Trojan can be activated by a specific event (e.g., specific input sequence, internal data value, system state) or by predefined timing (e.g., a specific number of clock cycles) [15, 16]. Second, the enable signal of the BFSK transmitter is provided outside the salware: it is the same signal as that used to activate the IP identification. For malware, the BFSK transmitter's enable signal is driven internally by the hardware Trojan control logic. In this case, the Trojan activates the enable signal when it is ready to send the stolen data. Note that an enable signal is required in both applications to reduce the power consumed by the ring oscillator.

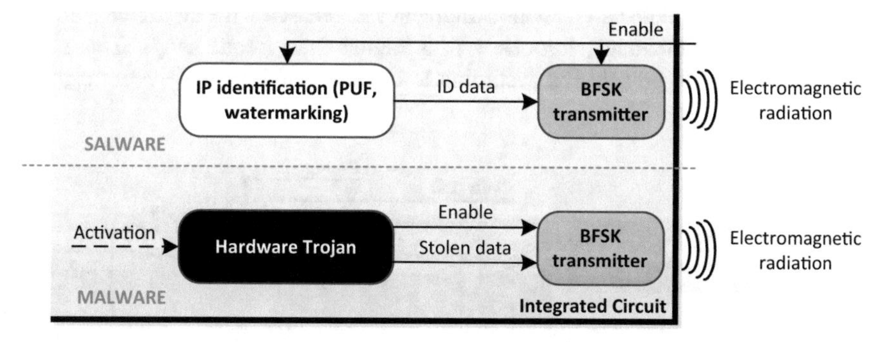

Fig. 5.6 Electromagnetic transmission of data (i.e., IP identification data or stolen secret data by a hardware Trojan such as the secret key for symmetric cipher)

Moreover, a permanently activated transmitter could be detected more easily by a spectral analysis of electromagnetic emanations of the device and could also cause local heating and premature aging of the chip.

5.3.2 Ultra-Lightweight Digital BFSK Transmitter

Electromagnetic radiation is an efficient side channel since, unlike measurement of power consumption, electromagnetic radiation can be measured locally and contactless. For this reason, we use the electromagnetic channel for our IP identification scheme. To this end, we designed an ultra-lightweight BFSK transmitter which could be activated outside the device by an ID checker or internally by a specific event (e.g., specific input sequence, internal data value, system state). Note that an enable signal is required to reduce the power consumed by the ring oscillator. Moreover, a permanently activated transmitter could be detected more easily by a spectral analysis of electromagnetic emanations of the device and could also cause local heating and premature aging of the chip.

BFSK is one of the common modulation schemes used in digital communication. The binary data are sent using a sinusoidal carrier at two frequency tones f_0 and f_1, representing high ('1') and low ('0') logic levels. The binary data arriving at the transmitter input at certain bitrates determine the commutation of the tones at the transmitter output. The proposed BFSK transmitter uses a dedicated configurable ring oscillator, as shown in Fig. 5.7. The configurable ring oscillator is designed using one multiplexor, $N + K$ delay elements, and a feedback chain controlled by a NAND gate for activation of transmission to reduce power consumption. Actually, the transmitter is used only during a short time when the enable signal is high, and it consumes power only during this small piece of time. The power consumption of this transmitter is thus completely negligible.

Input data controls the multiplexor, as shown in Fig. 5.7. When input data is low, the ring oscillator uses N delays and its oscillation frequency is f_0. When input data is high, the ring oscillator uses $N + K$ delays and its oscillation frequency is f_1. Since the ring oscillator's oscillation frequency decreases with an increase in the number of delay elements, frequency f_0 is higher than frequency f_1. These two frequencies have to be selected according to the bandwidth of the electromagnetic

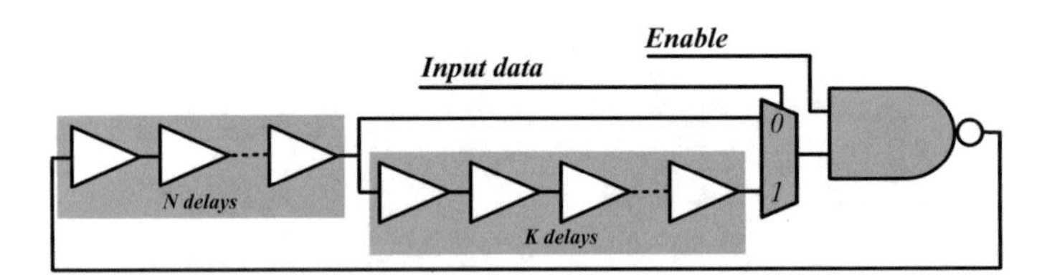

Fig. 5.7 Architecture of the ultra-lightweight digital BFSK transmitter based on a configurable ring oscillator

analysis platform, which is used to acquire and measure the transmitted signal. The bandwidth of our test bench, which is described in the Sect. 5.3.3 of this chapter, was limited to 100 MHz and 1 GHz by the low-noise amplifier.

The proposed BFSK transmitter was first implemented in Microsemi FUSION flash based FPGA (130 nm CMOS technology) containing 600K logic gates (M7AFS600). The device contains 13824 tiles, each tile can be used to implement one D-flip-flop or one configurable multiplexor-based logic block implementing any 3-input logic function.

The configurable number of delays in the ring oscillator of the proposed BFSK transmitter makes it possible to select precisely the two frequencies f_0 and f_1 using parameters N and K. Table 5.3 lists the ring oscillator frequencies and the number of Fusion tiles used by the BFSK transmitter for five values of N and K, with N ranging from 0 to 4, and K ranging from 1 to 5. According to Table 5.3, f_0 can be chosen between 119 MHz ($N = 4$) and 385 MHz ($N = 0$) and f_1 can be chosen between 70 MHz ($N = 4$, $K = 5$) and 280 MHz ($N = 0$, $K = 1$). The exact value

Table 5.3 Hardware implementation results of the BFSK transmitter

N	K	F_0 (MHz)	F_1 (MHz)	Fusion tiles	LUT4	EG
0	1	385	280	3	2	4.67
	2	383	210	4	3	5.34
	3	384	151	5	4	6.01
	4	385	130	6	5	6.68
	5	381	111	7	6	7.35
1	1	272	189	4	3	5.34
	2	272	156	5	4	6.01
	3	270	120	6	5	6.68
	4	271	106	7	6	7.35
	5	269	93	8	7	8.02
2	1	168	144	5	4	6.01
	2	169	124	6	5	6.68
	3	169	100	7	6	7.35
	4	168	91	8	7	8.02
	5	168	79	9	8	8.69
3	1	146	128	6	5	6.68
	2	147	112	7	4	7.35
	3	146	92	8	5	8.02
	4	145	84	9	6	8.69
	5	144	74	10	7	9.36
4	1	123	110	7	6	7.35
	2	121	98	8	7	8.02
	3	122	83	9	8	8.69
	4	121	77	10	9	9.36
	5	119	70	11	10	10.03

of f_0 depends on the number of delay elements, but also on the placement and routing of the transmitter. For the values N and K listed in Table 5.3, the frequency variation was less than 1.7 % (the maximum frequency deviation in Table 5.3 is 2 MHz when $N = 4$).

The number of tiles used by the BFSK transmitter is very low, i.e., from 3 tiles ($N = 0$, $K = 1$) to 11 tiles ($N = 4$, $K = 5$). These values are equivalent to less than 0.022 % and less than 0.080 % of the total number of tiles included in the targeted 600K-gate FUSION FPGA, respectively. This very small number of tiles is very promising for good dissimulation of the BFSK transmitter inside the sea of gates/tiles. The number of FUSION tiles required by the BFSK transmitter is given by the following Eq. (5.5).

$$Number\_FTiles = N + K + 2 \tag{5.5}$$

In order to estimate the number of resources needed for implementation with Xilinx SRAM FPGA or Altera SRAM FPGA, Table 5.3 gives the number of 4-input look-up-tables (LUT4) used by the BFSK transmitter with such FPGAs. The number of LUT4 required by the BFSK transmitter is given by the following Eq. (5.6).

$$Number\_LUT4 = N + K + 1 \tag{5.6}$$

To evaluate the logical resources needed by the BFSK transmitter in ASIC implementations, the right-hand column in Table 5.3 gives the number of equivalent gates (EG) in the transmitter. The gate count was estimated using the Virtual Silicon standard cell library based on the UMC L180 0.18 μm 1P6 M Logic process (UMCL18G212T3 [17]). The delay gates are replaced by more efficient standard NOT gates. The gate count of a standard NOT gate is 0.67 EG, and that of the standard multiplexor, 2.33 EG. The standard NAND gate uses 1 EG. So the number of gates of the whole BFSK transmitter ranges from 4.67 EG ($N = 0$, $K = 1$) to 10.03 EG ($N = 4$, $K = 5$). Note that one flip-flop requires between 5.33 EG and 12.33 EG to store a single bit [17].

Such a transmitter is clearly ultra-lightweight in both FPGA and ASIC implementations. The small logical resources requirement of the proposed spy circuitry makes reverse engineering it harder, although not impossible [18]. Even with recent CMOS technologies, the attacker can reverse engineer ICs using a scanning electron microscope and an automatic tool for circuitry extraction [18, 19]. Nevertheless, the smaller the piece of hardware used for BFSK transmitter the harder it is to detect during reverse engineering. Detection of the transmitter using standard Trojan detection methods [20, 21] is not feasible because the transmitter does not change the data path of the circuit and because of the ultra-low signal-to-noise ratio on the electromagnetic channel, as shown in our experimental results below (see the Sect. 5.3.3 of this chapter).

5.3.3 Experimental Results

The electromagnetic radiation of the device was evaluated using the near-field electromagnetic analysis test bench described in [14]. The border between the far field and the near field can be considered to be about 23 mm from the device, depending on the hardware concerned. The most important part of the test bench is the acquisition chain. It determines the signal-to-noise ratio and measurement precision.

The chain, as presented Fig. 5.8, is composed of:

- A Langer magnetic probe with a frequency range of from 30 MHz to 3 GHz and a spatial resolution of approximately 500 μm.
- A Miteq low-noise amplifier with a frequency range of from 100 MHz to 1 GHz.
- A Tektronix real-time signal analyzers RSA5106B with a frequency range from 1 Hz to 6.2 GHz [22].

As presented in Fig. 5.8, the device to be tested (the board) is fixed to a XYZ table with repeatability of movement of 1 μm. The test bench, including the acquisition chain, XYZ table, FPGA configuration and power supply variations, is controlled by a computer. This test bench was first developed for electromagnetic attacks of TRNGs [13], and [14].

The targeted FPGA for the experimental work is an Altera Cyclone III EP3C25 that uses a 65 nm CMOS technology. It contains 24624 four-inputs LUT and 608256 RAM bits.

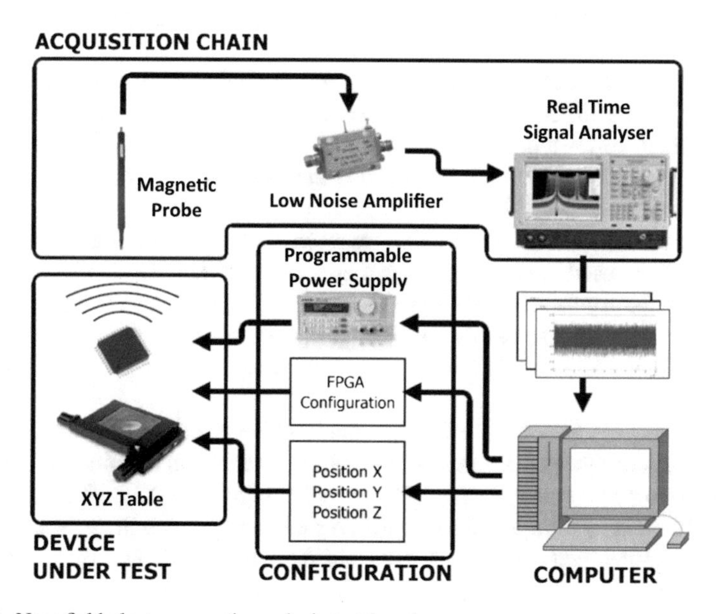

Fig. 5.8 Near-field electromagnetic analysis test bench

Electromagnetic analysis of IC is contactless, local, and can be spatial or/and temporal. This last point makes it possible to perform frequency analysis of the electromagnetic emanation. In the your bench the spectral range is limited to 100 MHz and 1 GHz. Standard devices aimed at direct BFSK demodulation cannot be used for these relatively high frequencies. Available integrated BFSK demodulators are limited to a few dozen megahertz. For this reason, we developed a dedicated BFSK demodulation scheme for our needs, in which a spectral analysis of the low-noise amplifier output (a component of the test bench) is performed to measure the f_0 and f_1 spectral contribution. The transmitted high (low) level is detected when f_1 spectral contribution is higher (lower) than that of f_0.

For the coherent demodulation of the electromagnetic radiation, we propose a slippery window spectral analysis. Indeed, overall spectral analysis masks the effects of the no stationarity of the signal and therefore provides no information about its temporal evolution. Slippery window spectral analysis is a three-dimensional representation of the signal: amplitude, frequency, and time. It requires two quantities *Fw*, the width of the FFT window frame and the difference $\Delta\tau$ between two frames. For our experiment, we chose *Fw* equal to 16384 points (2^{14}-point FFT) and $\Delta\tau$ equal to 100 points. For each frame, the FFT provides the software demodulator with the amplitude of signals f_0 and f_1 which enables the demodulator to distinguish between a transmitted '1' or '0.'

To illustrate data transmission from the circuit via the EM channel, we used a shift register that stored the following 16-bit sequence: "0101000111110011." The clock frequency of the shift register is 1 MHz. When the enable signal of the transmitter is given, the sequence is sent cyclically to the BFSK transmitter, which transmits it via the electromagnetic channel. The following gives the result of the coherent demodulation obtained at a 1 Mbps bit rate, which served as a proof of concept.

Figures 5.9 and 5.10 present the temporal evolution of the spectral analysis (amplitude) of the BFSK transmitter's electromagnetic emission when $N = 6$ and $K = 10$, which corresponds to the following frequencies: $f_0 = 289$ MHz (Fig. 5.9) and $f_1 = 119$ MHz (Fig. 5.10). Notice also that we placed a small antenna in the close vicinity of the ring. With $N = 6$ and $K = 10$ the BFSK transmitter uses only 17 four-inputs LUT of the FPGA that represents 0.065 % of the available logical resources of the used Altera FPGA for theses experimental results.

The used Tektronix real-time signal analyzers [22] allows us to obtain spectral cartography with direct reading of the patent that contains the transmitted data sequence. Figure 5.11 shows the spectral cartographies obtained at $f_0 = 289$ MHz and $f_1 = 119$ MHz.

Without knowledge of the BFSK parameters, the electromagnetic transmission cannot be easily detected because it cannot be distinguished from spectral noise. The signal-to-noise ratio of the BFSK transmission is -135 dB for a 1 GHz bandwidth. Such an ultra-low SNR represents efficient protection against unwanted BFSK transmitter detection via a side channel. However, knowing the *N* and *K* parameters, the BFSK designer can calibrate the demodulation (determine the

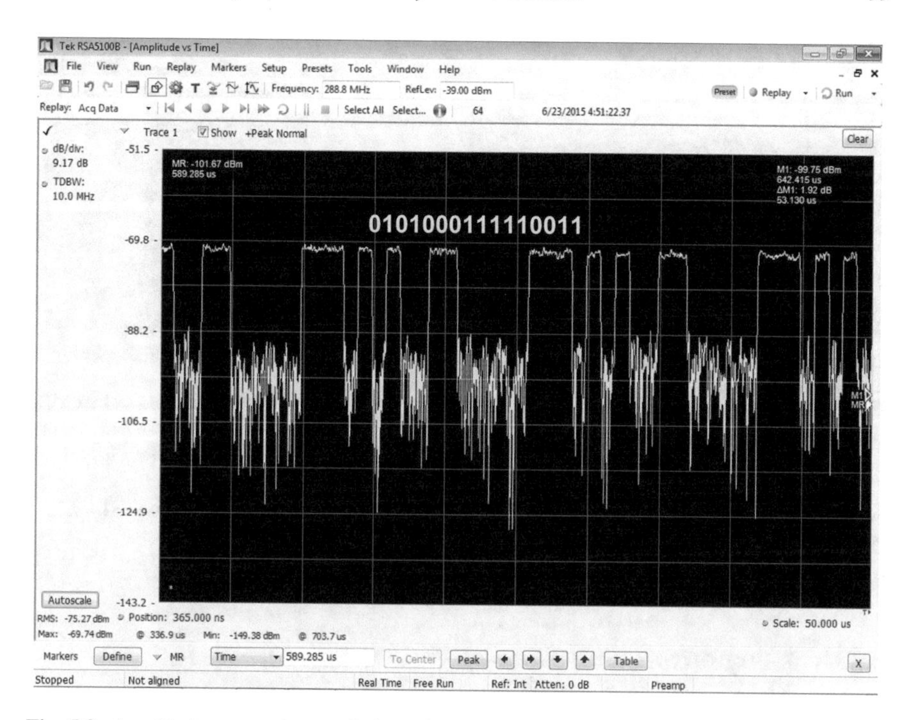

Fig. 5.9 Amplitude versus time evolution of the spectral analysis at $f_0 = 289$ MHz

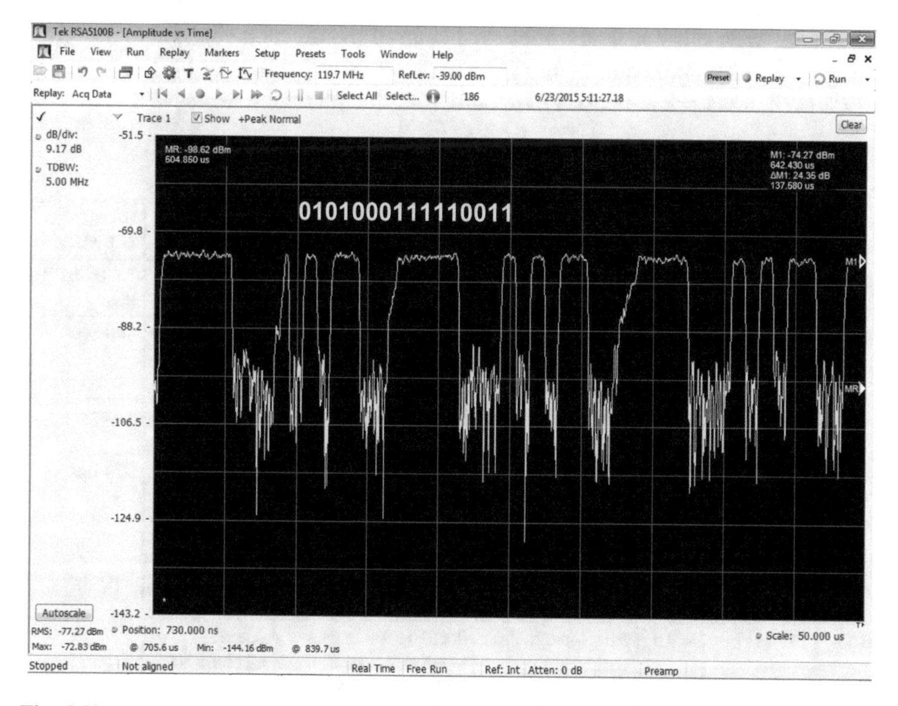

Fig. 5.10 Amplitude versus time evolution of the spectral analysis at $f_0 = 119$ MHz

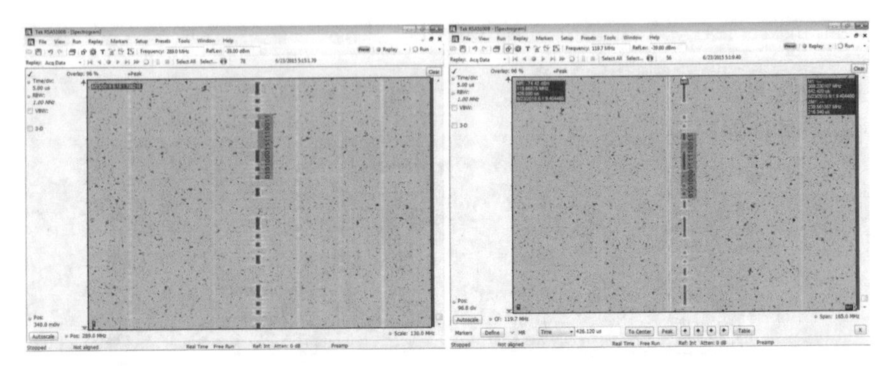

Fig. 5.11 Spectral cartographies center (*red trace*) on f_0 = 289 MHz (*left*) and on f_1 = 119 MHz (*right*) with 1 Mbps data rate

two frequencies) by electromagnetic analysis of the ring oscillators based on the differential spectral analysis as described in [14].

5.3.4 Comparison with State-of-the-Art Spy Circuitries Using a Side Channel

Table 5.4 compares the implementation of the proposed ultra-lightweight BFSK transmitter with other recently published state of the art methods. Table 5.4 gives the spy circuitry application (*App.*) for each reference; this may be IP protection (*IPP*) or hardware Trojan (*HT*) or both (for the presented work, ***PW*** [12]). In

Table 5.4 Summary of characteristics of spy circuitries exploiting side channels

App.	References	YoP	Side channel	Hardware resources	Target	Bit rate @1 MHz
IPP	[7]	2008	Thermal emanation	255 Spartan-3 slices	Xilinx Spartan-3	14×10^{-3} bps
	[8]	2008	Power consumption	16 * 16-bit circular shift registers	Xilinx Spartan-3 and Virtex-II	400 bps
	[9]	2010	Power consumption	16-bit circular shift-register	Xilinx Virtex-II Pro	1 kbps
HT	[23]	2009	Power consumption	8 parallel D-flip-flops or 16-bit circular shift register	Xilinx Spartan-3E and Virtex-II Pro	970 bps
	[25]	2013	Power consumption	16-bit circular shift registers per bit	Xilinx Virtex-5	1.9 kbps
Both	PW [11]	2015	Electromagnetic emanation	1 configurable ring oscillator (like a D-flip-flop in ASIC)	Altera Cyclone III	1 Mbps

addition, Table 5.4 gives the year of publication (*YoP*), the side channel used, the hardware resources required only for the leakage generator (for example, we do not take the hardware used for IP watermark generation or the Trojan's payload into account). Unfortunately, the principles compared do not use the same hardware. For the sake of correctness, we give the implementation results as they are presented in the referenced papers. Nevertheless, the implementation bitrate of these previously published works can be roughly compared with our proposed solution. Based on published data, we computed the bitrate of all the proposals by using the number of clock cycles needed to send information via the side channel. For all the references presented in this table, the bitrate was computed using a 1 MHz frequency for data synchronization (same frequency is used during the experimental works presented previously).

As can be seen in Table 5.4, the proposed work reaches the highest bitrate. The reason for such a good performance is first that we use a spectral analysis of the local electromagnetic leak based on a simple frequency modulation. Except for [7], all the other solutions use a global measurement of power consumption, which reduces the signal-to-noise ratio of the information leaked via the side channel. Our proposal is clearly the smallest spy circuitry ever published. Although solutions based on circular shift registers are well adapted to last generation FPGA families, since the 16-bit shift registers can be designed using only one look-up table, they are not suitable for ASIC technologies. Currently, an ASIC implementation of a 16-bit shift register requires 16 flip-flops whereas the solution we propose occupies an area equivalent to only one D-flip-flop.

In this chapter, we present the proposed spy circuitry for IP protection, but it can also be used for hardware Trojan. Most of the other proposals could also be used for both applications. Note that in 2012, Kasper et al. proposed to use the work initiated in [23] for hardware Trojan or IP watermarking implementation [24]. However, by using electromagnetic emanation and a configurable ring oscillator, the proposed solution is the most convincing for industrial applications (e.g., those aimed at IP protection) because of its very small area and high bitrate.

5.3.5 *Industrial Scenarios Using the Proposed P Protection*

According to the previous section, in comparison with other works, our propose goes clearly toward using a spy circuit in an industrial context for IP protection. Two industrial scenarios are presented in the following. The first scenario is the identification of embedded IP in the supply chain. This identification is used in order to be sure to do not use counterfeiting (fake) devices.

It is therefore crucial and strategic to be able to detect counterfeit IC as soon as possible in the supply chain (this is particularly crucial for military and space grad devices). Figure 5.12 shows a possible framework to manage the device under test (control the enable signal) and check the IP identification by using an EM probe, an amplifier and a dedicated acquisition system including a spectral analysis and the

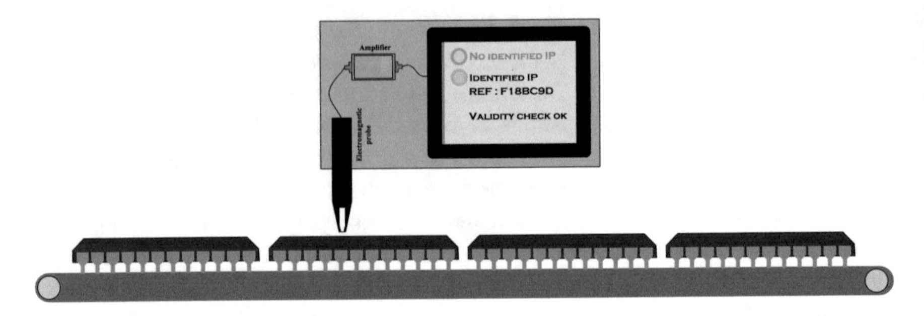

Fig. 5.12 Rapid and contactless IP identification in the supply chain by using EM transmission of IP' ID

proposed demodulation method. Due to the high bit rate of the proposition solution the identification of the ID requires less than 500 μs (with 1 Mbps bit rate). This counterfeiting detection could be completed by other physical (invasive or not) and optical inspection [26].

The second scenario occurs when an IP designers would like to verify the illegal presence of its IP inside a device (ASIC or FPGA). In this case the proposed transmitter provides to the identification scheme a data like a PUF [27] or a watermarking. Watermarking is a technique of steganography which provides the ownership of an IC (or an IP) by checking for the presence of hidden information called the watermark [2, 3]. Most of the watermarking methods proposed in the literature need a complex verification scheme. Nevertheless it is possible to use power consumption as proposed in [2] but it is easy and cheap to use global countermeasure in order to mask the power consumption due to the watermark [28]. Using electromagnetic emanation in this scenario is better because as electromagnetic emanation is local it is really hard to mask it by using a global countermeasure. Moreover, in this paper we have shown that due to the SNR of BFSK signal, it is unrealistic to try to detect it without the precise knowledge of the used frequencies for data transmission.

5.4 Conclusion

IP protection has become crucial topics for hardware security due to the lack of trust in IP market. In this chapter we have first presented an IP watermarking verification scheme that exploits the power consumption of an integrated circuit. Experimental results are presented and prove that it is possible to clearly identify different FSMs with the same watermark key (*Kw*) and the same FSM with a different watermark key too. Thus, our method is robust against some kinds of collisions. In addition, the verification scheme is insensitive to the CMOS process variation. Then, we have presented an ultra-lightweight transmitter of IP identity using the electromagnetic side channel. Based on a configurable ring oscillator, our solution exploits a BFSK signal to transmit information by way of the electromagnetic channel. By performing a slippery window spectral analysis of the near-field electromagnetic emanations captured locally over the BFSK transmitter circuitry, the proposed transmission achieves a high bitrate (experimentally at less 1 Mbps), which has not been achieved before. Moreover, the transmitter occupies very small area less than the requirement of a small D-flip-flop. Such a small requirement of logical resources makes reverse engineering of the chip very difficult and detection of the transmitter using standard Trojan detection methods is not feasible.

Acknowledgment The work presented in this paper was realized in the frame of the SALWARE project number ANR-13-JS03-0003 supported by the French "Agence Nationale de la Recherche" and by the French "Fondation de Recherche pour l'Aéronautique et l'Espace."

References

1. B. Colombier, L. Bossuet, Survey of hardware protection of design data for integrated circuits and intellectual properties. IET Comput. Digital Tech. **8**(6), 274–287 (2014)
2. B. Legal, L. Bossuet, Automatic low-cost IP watermarking technique based on output mark insertion. J. Des. Autom. Embedded Syst. **16**(2), 71–92 (2012). Springer
3. C. Marchand, L. Bossuet, E. Jung, IP watermark verification based on power consumption analysis, in *Proceedings of the 27th IEEE International System-on-Chip Conference, SOCC 2014* (2014), pp. 330–335
4. P. Kocher, J. Jaffe, B. Jun, Differential power analysis, in M. Wiener (ed.), *Proceedings of the 19th Annual International Cryptology Conference, CRYPTO 1999*. Lecture Notes on Computer Science, vol. 1666 (Springer, 1999) pp. 388–397
5. N. Kamoun, L. Bossuet, A. Gazel, Experimental Implementation of 2ODPA attacks on AES design with flash-based FPGA Technology, in *proceedings of the 22nd IEEE International Conference on Microelectronics*, IMC 2010, pp. 407–410
6. L. Bossuet, D. Hely, SALWARE: Salutary hardware to design trusted IC, in *Workshop on Trustworthy Manufacturing and Utilization of Secure Devices, TRUDEVICE 2013* (2013)
7. C. Marsh, T. Kean, D. Mclaren, Protecting designs with a passive thermal tag, in *Proceedings of the 15th IEEE International Conference on Electronics, Circuits and Systems, ICECS 2008* (2008), pp. 218–221
8. D. Ziener, J. Teich, Power signature watermarking of IP cores for FPGAs. J. Signal Process. Syst. **51**, 123–136 (2008). Springer
9. G.T. Becker, M. Kasper, A. Moradi C. Paar, Side-channel based watermarks for integrated circuits, in *Proceedings of the IEEE International Symposium on Hardware-Oriented Security and Trust, HOST 2010* (2010), pp. 30–35
10. S. Kerckhof, F. Durvaux, F.X. Standaert, B. Gérard, Intellectual property protection for FPGA designs with soft physical hash functions: first experimental results, in *Proceedings of the IEEE International Symposium on Hardware-Oriented Security and Trust, HOST 2013* (2013), pp. 7–12
11. S. Katzenbeisser, Ü. Kocabaş, V. Rožić, A.R. Sadeghi, I. Verbauwhede, C. Wachsmann, PUFs: Myth, fact or busted? a security evaluation of physically unclonalble functions cast in silicon, in *Proceedings of the Workshop on Cryptographic Hardware and Embedded Systems, CHES 2012*, Lecture Notes on Computer Science, vol. 7428 (Springer, 2012), pp. 283–301
12. L. Bossuet, P. Bayon, V. Fischer, Contactless transmission of intellectual property data to protect FPGAs designs, in *Proceedings of the IFIP/IEEE International Conference on Very Large Scale Integration, VLSI-SOC 2015* (2015), pp. 19–24

13. P. Bayon, L. Bossuet, A. Aubert, V. Fischer, F. Poucheret, B. Robisson P. Maurine, Contactless electromagnetic active attack on ring oscillator based true random number generator, in *Proceedings on International Workshop on Constructive Side-Channel Analysis and Secure Design, COSADE 2012*, Lecture Notes in Computer Science, vol. 7275 (Springer, 2012), pp. 151–166
14. P. Bayon, L. Bossuet, A. Aubert, V. Fischer, EM leakage analysis on true random number generator: frequency and localization retrieval method, in *Proceedings of the Asia Pacific International Symposium and Exhibition on Electromagnetic Compatibility, APEMC 2013* (2013)
15. R. Karri, J. Rajendran, K. Rosenfeld, M. Tehranipoor, Trustworthy hardware: identifying and classifying hardware trojans. IEEE Comput. **43**(10), 39–46 (2010)
16. M. Tehranipoor, F. Koushanfar, A survey of hardware trojan taxonomy and detection, IEEE Des. Test Comput. **27**(1), 10–25 (2010)
17. Virtual Silicon Inc. 0.18 µm VIP Standard Cell Library Tape Out Ready, Part Number: UMCL18G212T3, Process: UMC Logic 0.18 µm Generic II Technology: 0.18 µm, 2004
18. R. Torrance, D. James, The state-of-the-art in semiconductor reverse engineering, in *Proceedings of the 48th Design Automation Conference, DAC 2011*, (ACM/EDAC/IEEE, 2011), pp. 333–338
19. P. Subramanyan, N. Tsiskaridze, W. Li, A. Gascon, W. Tan, A. Tiwari, N. Shankar, S. Seshia, S. Malik, Reverse engineering digital circuits using structural and functional analyses, in IEEE Trans. Emerg. Top. Comput. (2013)
20. D. Agrawal, S. Baktir, D. Karakoyunlu, P. Rohatgi, B Sunar, Trojan detection using IC fingerprinting, in *Proceedings of the IEEE Symposium on Security and Privacy* (2007), pp. 296–310
21. Y. Jin, Y. Makris, Hardware trojan detection using path delay fingerprint, in *IEEE International Workshop on Hardware-Oriented Security and Trust, HOST 2008* (2008), pp. 51–57
22. Tektronix, RSA5000 Series, Spectrum Analyzers Datasheet (2015). http://www.tek.com/sites/tek.com/files/media/media/resources/RSA5000-Series-Spectrum-Analyzers-Datasheet-37W2627414_1.pdf
23. L. Lin, M. Kasper, T. Güneysu, C. Paar, W. Burleson, Trojan side-channels: lightweight hardware trojans through side-channel engineering, in *Proceedings of Workshop on Cryptographic Hardware and Embedded Systems, CHES 2009*, Lecture Notes in Computer Science, vol. 5747 (Springer, 2009), pp. 382–395
24. M. Kasper, A. Moradi, G.T. Becker, O. Mischke, T. Güneysu, C. Paar, W. Burleson, Side channels as building blocks. J. Cryptogr. Eng. **2**(3), 143–159 (2012). Springer
25. S. Kutzner, A. Poschmann, M. Stöttinger, TROJANUS: An ultra-lightweight side-channel leakage generator for FPGAs, in *Proceedings of International Conference on Field-Programmable Technology, ICFPT 2013 (2013)*, pp. 160–167
26. M. Tehranipoor, U. Guin, D. Forte, Counterfeit Integrated Circuits—Detection and Avoidance (Springer, 2015)
27. L. Bossuet, X.T. Ngo, Z. Cherif, V. Fischer, A PUF based on a transient effect ring oscillator and insensitive to locking Durvaux, F.X. Standaert, and B. Gérard phenomenon. IEEE Trans. Emerg. Top. Comput. **2**(1), 30–36 (2014)
28. N. Kamoun, L. Bossuet, A. Ghazel, Correlated power noise generator as a low cost DPA countermeasure to secure hardware AES cipher, in *Proceedings of the International Conference on Signals, Circuits and Systems, SCS 2009* (2009), pp. 1–6

Chapter 6
Hardware Obfuscation: Techniques and Open Challenges

Georg T. Becker, Marc Fyrbiak and Christian Kison

6.1 Introduction

There are many applications for IC reverse-engineering. While there are legitimate reasons for IC reverse-engineering, some have malicious intend such as IP infringement and technological espionage. Particularly, Intellectual Property (IP) theft and counterfeit products are a major challenge for the industry. In many cases, the initial step in counterfeiting or stealing of IP is to reverse-engineer a chip or IP core in order to integrate the IP into one's own design illegitimately. Hence, there are various reasons why hardware companies demands obfuscation methods to hamper reverse-engineering of their designs. For security-critical devices reverse-engineering can also be a potential attack vector. An adversary can leverage reverse-engineering to disclose internal details of the design in order to enable further attacks on the system. For example, implementation attacks such as side-channel or fault attacks exploit implementation structures and thus an attacker gaining knowledge of the used implementation and countermeasures gains a significant attack advantage. In addition to these malicious goals, reverse-engineering can also be used to detect patent infringements and IP theft as well as to identify Hardware Trojans.

As a consequence, hardware obfuscation techniques that hamper reverse-engineering are of great interest. In this chapter, we present and discuss state-of-the-art hardware obfuscation techniques at two distinct levels. Hardware obfuscation at the layout level targets the extraction of the device's netlist. To be more precise, the underlying principle is to prevent the distinct identification of combinatorial gates. In Sect. 6.2, we provide a summary of the proposed layout-level obfuscation techniques and additionally a security evaluation. However, not every case of IP piracy starts with reverse-engineering of the targeted Integrated Circuit (IC). For example, most IP provides do not manufacture their own chips, but only sell IP cores in the

G.T. Becker (✉) · M. Fyrbiak · C. Kison
Department of Electrical Engineering and Information Technology,
Ruhr Universität Bochum, Universitätsstr. 150, 44780 Bochum, Germany
e-mail: georg.becker@ruhr-uni-bochum.de

L. Bossuet and L. Torres (eds.), *Foundations of Hardware IP Protection*,
DOI 10.1007/978-3-319-50380-6_6

form of hard and soft IP cores. In these scenarios, the adversary is already in possession of the netlist (without the need of IC reverse-engineering).

In order to prevent the disclosure of internal details, obfuscation transformations at the netlist level are required. In Sect. 6.3, we present and discuss the state-of-the-art proposed netlist-level obfuscation methods and automatic reverse-engineering capabilities. Furthermore, we address various limitations and open challenges for the different netlist-level obfuscation techniques.

6.2 Layout-Level Obfuscation

The first step in reverse-engineering of a targeted Application Specific Integrated Circuit (ASIC) is to obtain precise images of each chip's layer and subsequently to identify the individual gates and their connectivity. Based on this information, the netlist of the design can be derived which enables the reverse-engineer to analyze the design as well as to make copies of it. Hence, the principal goal of layout-level obfuscation is to hamper this netlist disclosure by the use of special combinatorial gates that cannot be correctly identified via visual reverse-engineering techniques by using a Scanning Electron Microscope (SEM). Layout-level obfuscation has been of interest in the industry for many years—the first patents date back to the 1980s [1]. However, despite the industry's ongoing interest in this topic, it has been largely ignored by the the scientific community and only recently the first works have appeared [2–5].

In the following, we introduce several proposals of how layout-level obfuscation can be realized. Particularly, we illustrate the so-called *camouflage gates* or *look-alike gates* that are utilized instead of standard cells in order to prevent the visual identification of the implemented logic function. These camouflage gates are the main building block in layout-level hardware obfuscation.

6.2.1 *Camouflage Gates*

The main idea of camouflage gates is to hide the logic function of the gates in design layers that are hard to detect visually. The main assumption of most camouflage gates is that metal layers and polysilicon layers are easily recognizable using visual reverse-engineering techniques such as SEMs. Hence, the goal of building camouflage gates is to create gates that are identical at these layers but still have a different functionality. The most common way to achieve this is to only change the dopant masks, i.e., to build gates that are identical on all design layers and only differ in the dopant polarity in some active areas. For example, this technique is used in [2, 3, 5]. A different approach is to use a mixture of real and dummy contacts to camouflage the functionality of the obfuscated gates [4]. The main idea behind this is that extra effort is needed to reverse-engineer the contacts. However, compared to dopant-based obfuscation, contact-based obfuscation is considerably easier to reverse-engineer. A more detailed analysis of the difficulty of reverse-engineering both dopant-based and contact-based obfuscation—as well as no obfuscation at all—is provided in Sect. 6.2.4.

Obfusgate camouflage gates: In the following, the main idea behind dopant-based camouflage gates is explained using the example of a camouflage gate called *Obfusgate*, which has been proposed in [3]. The heart of the Obfusgate is a so-called *Obfuscell*. Depending on the dopant polarity in its active areas, an Obfuscell can be configured to be either an inverter, a buffer (input = output), or to output a constant '1' or '0'. Figure 6.2a shows the layout of an Obfuscell. It has three active areas A1, A2, and A3. The dopant polarity within these areas define the configuration. To create a buffer for example, the input needs to be connected to the output via A2 as illustrated in Fig. 6.2b. For example, the active area A2 can be used to build a direct connection between the input and output. This is achieved by doping the entire active area A2 positively. On the other hand, if the middle region of A2 is doped negatively, this creates a p-n-p junction and hence a diode in cut-off. This is depicted in Fig. 6.3a. Hence, using different dopant polarity in the active area, one can connect or disconnect inputs. All dopant-based camouflage gates are based on this main idea [2, 3, 5]. In the Obfuscell design, the active areas A1 and A3 have the layout of a pmos and nmos transistor respectively (see Fig. 6.3b) for the doping of a pmos transistor) and hence can also create connections to the output. Figure 6.2b shows the four different possible configurations of the Obfuscell. An inverter is configured using A1 and A3 as pmos and nmos transistors respectively and disabling the connection in A2 as described above. If the Obfuscell is connected as a buffer, area A2 is doped positively and hence a connection is formed. Simultaneously, the two transistors in A1 and A3 needs to be "disabled", i.e., their outputs should be floating. How this can be done is depicted in Fig. 6.3d at the example a pmos transistor (area A1). The source contact which is connected to VDD is doped negatively instead of positively which basically creates a well-contact and a n-p junction to the output, i.e., a diode in cut-off. Hence, the output (drain) is floating and therefore the transistor is "disabled". Similarly, Fig. 6.3c shows how the drain can be connected to the input (connected to VDD) by doping the entire region positively. This way the active area A1 can be used to set the output to a constant one as needed for the "always 1" configuration (see Fig. 6.2b for details).

Hence, depending on the dopant polarity in A1, A2, and A3, the Obfuscell is either an inverter, a buffer (i.e., input = output), a constant "1" or a constant "0". This Obfuscell is then used as a building block to build Obfusgates that form the obfuscated standard cell library. Figure 6.1 depicts the structure of an "Obfusgate". It consists of five "Obfuscells" and a four-input NAND gate. Depending on the configuration of the Obfuscells (i.e., on their dopant), the Obfusgate can implement many different logic functions. For example, configuring all Obfuscells as buffers results in a four-input AND gate. Configuring the four Obfuscells that are used as inputs as inverters on the other hand results in a four-input NOR gate. In total, 162 different configurations, each with a unique logic behavior, can be realized with an

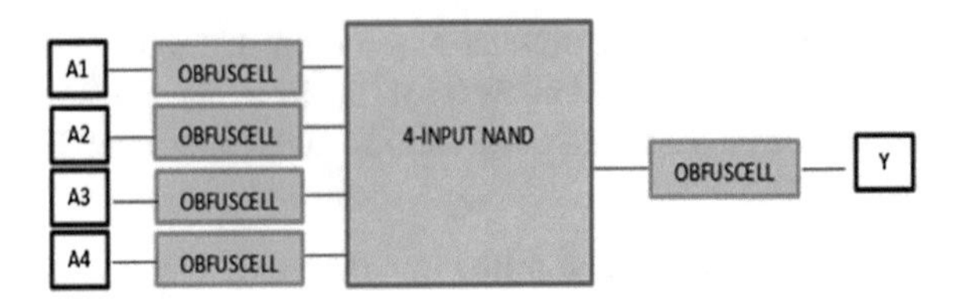

Fig. 6.1 Schematic of a single Obfusgate that consists of five Obfuscells together with a four-input NAND gate. Depending on the configuration of the Obfuscells, the Obfusgate can realize 162 different logic functions

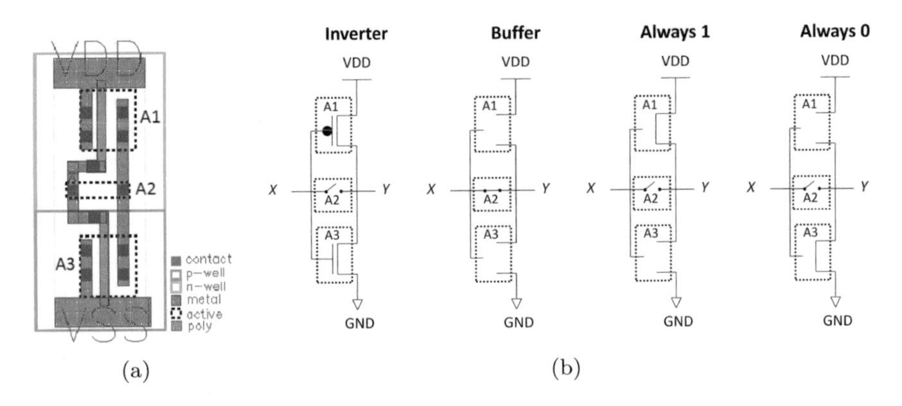

Fig. 6.2 **a** Layout view of the Obfuscell which has three active regions A1, A2, and A3 whose dopant polarity defines the logic function of the gate. The gate can be configured as a inverter, buffer, "always 1" or "always 0" gate as shown on the *right side* (**b**)

Obfusgate as depicted in Fig. 6.1. Furthermore, by setting an Obfuscell that is connected to an input of the Obfusgate to a constant "1", this input has effectively been turned into a "dummy input". Setting two input Obfuscells to a constant "1" and the other Obfuscells to a buffers results in an two input AND gate. The two inputs that are set to a constant "1" are the "dummy inputs" since the signal connected to these inputs has no effect on the output. Therefore, any signal can be connected to such an input, creating "dummy wires". Note that an attacker cannot distinguish between a dummy input and a regular input and hence this technique can increase the obfuscation significantly (Fig. 6.3).

DPD-LUT camouflage gates: Shiozaki et al. [5] used a different approach to build dopant-based camouflage gates. Their design is based on a 2-bit look-up table (LUT) similar to the LUTs used in FPGA designs. The input to these LUTs are special read-only memory cells called *Diffusion Programmable ROM* (DP-ROM) that are "programmed" using the dopant polarity. They function in exactly the same way as in the active area A2 of the Obfuscell and depicted in Fig. 6.3a. Each DP-ROM cell consists of two of these active areas and depending on the configuration of the dopant the output of the cell is either connected to VDD or GND.

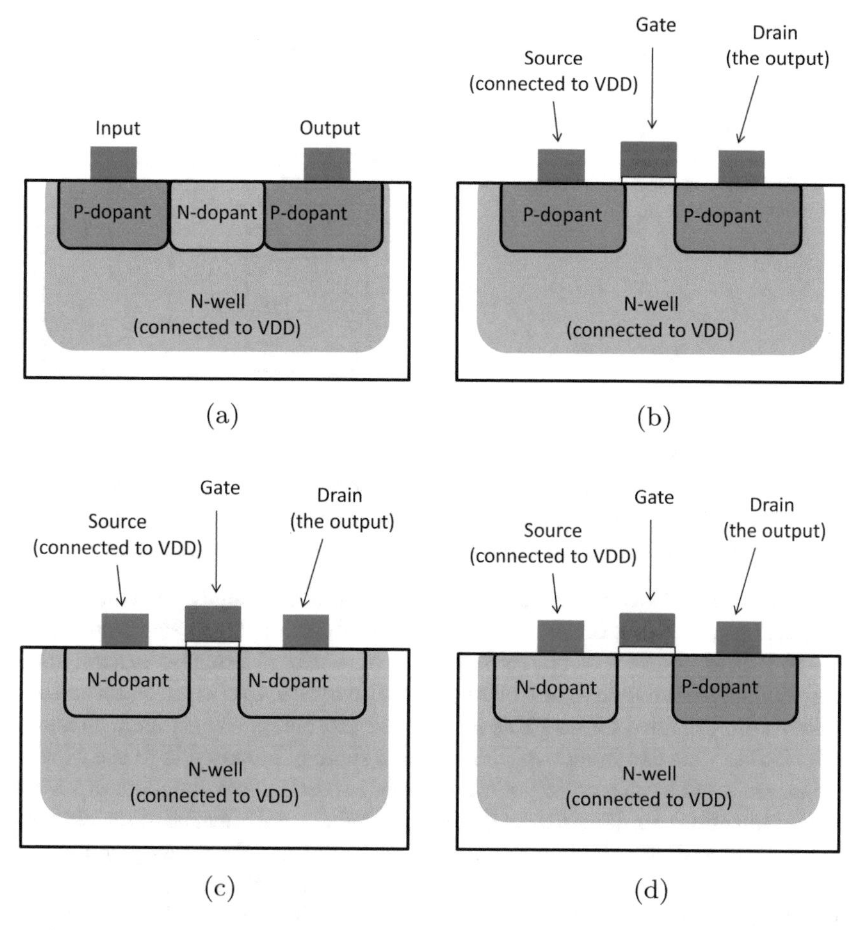

Fig. 6.3 Cross section view of the active areas. In **a** active area A2 is depicted configured in "cut-off", i.e., the input is disconnected from the output. By also doping the middle part positively the configuration is changed to a direct connection between the input and output. Active area A1 can be used as a pmos transistor as depicted in (**b**). In **c** the active area A1 is doped positively instead of negatively which results in a constant connection between input (connected to VDD) and the output. How a floating output can be realized for A1 is shown in (**d**). Only the source region of A1 is doped negatively which results in a n-p junction between source and drain, i.e., a diode in cut-off

The camouflage gate, which is called Diffusion Programmable Device Look-up Table (DPD-LUT), is depicted in Fig. 6.4. A DPD-LUT can be configured to any two-input logical function, i.e., it can realize $2^4 = 16$ different functions. We would also like to note that this design approach is not restricted to two-input LUTs. In an analogous manner, it is possible to build a three-input DPD-LUT or, as in FPGAs, four- or five-input LUTs. Using larger LUTs will likely result in a larger overhead but also in a higher grade of obfuscation. The smallest overhead could probably be achieved by combining DPD-LUTs of different sizes. However, the distribution

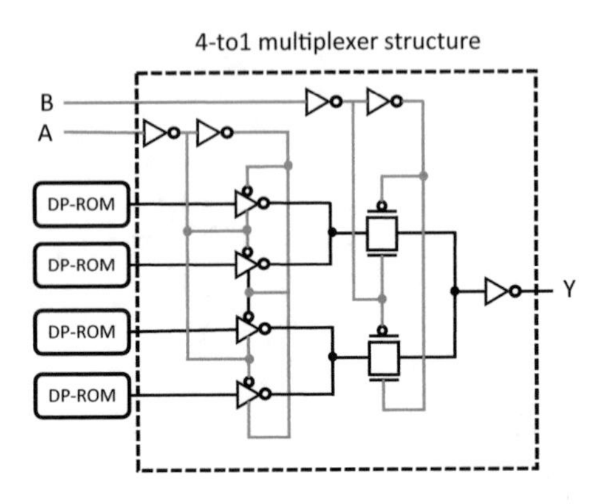

Fig. 6.4 The structure of a DPD-LUT. It consists of four DP-ROM cells and a two-input multiplexer structure with the inputs *A* and *B* and the output *Y*. Any 2-bit logical function can be implemented with such a DPD-LUT cell

of differently sized DPD-LUTs might help an attacker gain some insight into the obfuscated design. Questions like this have not been researched yet, and hence it is not clear to what extent the combination of different look-alike gates into one common obfuscated standard cell library can decrease obfuscation strength.

SMI's approach: Cocchi et al. proposed two different strategies to build camouflage gates [2]. The first one is to construct custom camouflage gates whose logic function is hard to reverse-engineer but which are easily identifiable as camouflage gates, similar to the approaches of Malik et al. and Shiozaki et al. Unfortunately, no details are provided on how this is achieved and how many different functions one look-alike gate can implement. The second strategy proposed is to use existing standard cells and to only modify a few in order to create a new functionality. Since these modifications are hard to detect, a reverse-engineer will mistake the camouflage gate for a "normal" standard cell and hence will come up with a faulty netlist. The advantage of this approach is that only a few camouflage gates might offer enough obfuscation in certain situations, while greatly reducing the introduced overhead. However, the disadvantage is that a single camouflage gate offers less obfuscation since it usually can only realize very few different logic functions. How exactly Cocchi et al. modified the standard cells and how many different functions such a cell can implement has not been disclosed. However, this approach shares many similarities with the dopant-level hardware Trojans presented at CHES 2013 [6], which also change the functionality of standard cells while making the detection of these modifications as hard as possible. Basically, the gates are modified as also done in the Obfuscell by connecting outputs to VDD or ground and removing transistors as depicted in Fig. 6.3. Hence, there is a large overlap in the construction of camouflage gates and stealthy layout-level hardware Trojans, and the Obfusgate design was inspired by these Trojans.

6.2.2 *Obfuscating the Connectivity:*

Camouflage gates obfuscate the logic function of individual gates. However, they do not conceal the connectivity, i.e., a reverse-engineer can still see which gates are connected with each other. Ideally, a reverse-engineer should not infer the function of a block after it has been obfuscated. But the connectivity of the individual gates reveals a lot of useful information to a reverse-engineer. This was illustrated in [3] with the example of the block cipher PRESENT [7]. The PRESENT round function is depicted in Fig. 6.5a. The key insight is that when camouflage gates are used, the logic of the gates are not known but. When grouping cells that are connected with each other together, the resulting graph would look like Fig. 6.5b. The white boxes represent blocks whose logic function is not known to the reverse-engineer due to the use of camouflage gates. However, since the connections are known, it is not difficult to, e.g., to identify the 4-bit SBoxes used in PRESENT, since they are functions with four inputs and exactly four outputs. The fact that four-input functions as, e.g., eight-input functions reveals a lot of information to a reverse-engineer about the employed encryption function.

Hence, since a lot of information is not obfuscated, camouflage gates by themselves are not enough to prevent reverse-engineering if the attacker's goal is to collect information about the design structure or to identify the location of certain IP blocks within a chip. In order to solve this problem, the Obfusgate library heavily uses "dummy wires" that conceal the connectivity. More than half of the obfuscation gates in the AES SBox and PRESENT round functions originally are two-input gates. Each Obfusgate that is configured as a two-input gate has two dummy inputs and hence also two dummy wires. In the proof-of-concept implementation of the Substitution and Permutation Layer of PRESENT, 941 dummy wires and 1103 normal wires are used [3] This very large amount of dummy wires effectively hides the connectivity and hence the structure of the design since an attacker cannot differentiate between dummy wires and real wires. However, it is important to note that

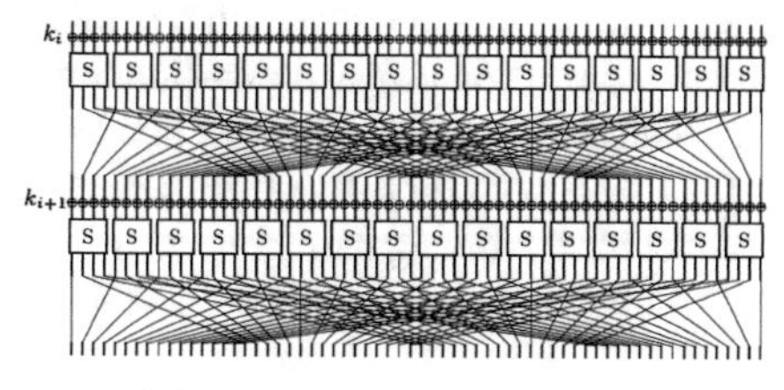

(a) PRESENT round function

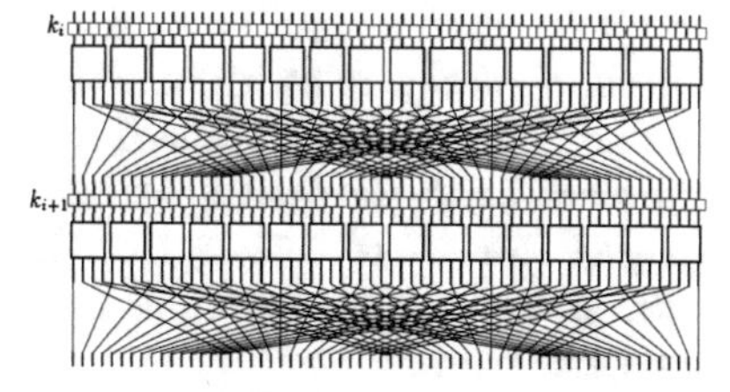

(b) PRESENT round function with obfuscated combinatorial gates

Fig. 6.5 **a** Figure of the PRESENT round function, taken from [7]. **b** When the combinatorial gates are replaced with camouflage gates, a reverse-engineer does not know the logic function of the SBoxes any longer. The structure of the round function on the other hand is still clearly visible due to the wires connecting the individual blocks and registers

this level of obfuscation comes with a large area overhead. Furthermore, the current version randomly connects the dummy wires. While this works for small designs such as the PRESENT round function, for larger designs the routing overhead would increase to a level that would make routing impossible. Hence, just random connections do not scale for large designs. Thus, how to efficiently obfuscate connectivity information at the layout level is an interesting open research problem. In general, the addition of dummy wires or connections can also be achieved using the camouflage gates proposed in [2, 4, 5]. This can be realized by inserting additional gates that only have the purpose of creating dummy wires. Again, the optimum number of additional gates and how to integrate them has not been analyzed and is therefore an open question.

6.2.3 *Further Obfuscation Techniques*

Besides using camouflage gates, other techniques to hamper reverse-engineering at the layout level have been proposed. For example, reverse-engineering non-volatile memory can be more difficult as reverse-engineering combinatorial memory [8]. The idea is therefore to not implement the entire design using normal combinatorial gates but also include non-volatile memory cells that are programmed after manufacturing. The content of these memory cells then determines the logic behavior of the chip. For example, this technique could be combined with DPD-LUT camouflage gates: Instead of using DP-ROM cells that are programmed based on dopant polarity, other non-volatile memory cells that are programmed after manufacturing can be used. One advantage of using non-volatile memory is that this also prevents the factory from over-producing the ICs. After manufacturing the fabricated chips are non-functional and hence only the IP owner can program and hence activate the chips.

Another technique which makes layout-level reverse-engineering is the use of special filler cells [2]. Typically, in a digital chip there often gaps between gates due to routing constrains, etc. These gabs are usually filled with so-called "filler cells" to fulfill certain design rules. These filler cells can be easily identified as non-functional cells during reverse-engineering. The idea is to instead use cells that look like legitimate gates, i.e., replace the non-functional filler cells with (non-functional) camouflage gates. Since a reverse-engineer does cannot easily distinguish these cells from functional camouflage gates, this can significantly increase the required reverse-engineering effort.

6.2.4 *Reverse-Engineering Camouflage Gates*

Ideally, camouflage gates make it impossible to reverse-engineer the gates using visual techniques by optical means. Having a process to determine a precise doping level and impurity is of outmost importance for the chip production and their failure analysis. Special processes are necessary to measure impurities and dopants. Therefore, the dopant-based and via-based camouflage gates do not prevent reverse-engineering in general, but rather hamper the process. In this section, we briefly discuss several reverse-engineering techniques that are able to reveal the functionality of the proposed camouflage gates. Notably these techniques emerged from failure analysis, trying to locate faults in dopant concentrations, defects, or impurities.

Delayering and Hardware Reverse-Engineering: The art of Hardware Reverse-engineering begins at the Printed Circuit Board (PCB) and package level of the IC piece of hardware. First the IC is cropped out or de-soldered from the PCB. Please note that this step is nontrivial for some flip-chip packages with underfill. The challenge to protect the die is becoming ever more difficult with reduced die size and thickness. Second, the package has to be removed by wet-chemical or mechanical means. Hereby, again, the die is to be protected from any harm which often results in choosing wet-chemical depackaging, as the die is protected by the seal-layer[1] from the front side. The backside offers enough silicon in the bulk to withstand carefully applied depackaging processes as well. The bonding wires are of special concern, as newer copper bondings are, compared to gold bonding wires, hard to preserve. For the invasive hardware reverse-engineering the wires can be neglected once their connectivity is known or the connectivity can be derived. Advanced techniques for finding bonding wire connectivity can be done by (3D) X-Ray or selective packaging delayering with a mill.

Once the die is fully recovered, the die is alternately delayered and digitalized by optical means or in a SEM/Focused Ion Beam (FIB). The following delayering processes are, again, a combination of different wet-chemical and mechanical polishing. Here it is of outmost importance for the quality of the process, to handle the equipment in an experienced way. Planarization of the current layer with a huge surface to thickness ratio is one, if not the hardest challenge to master. Please note knowing your Region of Interest (ROI) comes very handy at this point, as the planar surface can be reduced significantly. The reverse-engineer can pinpoint his ROI while neglecting the rest of the chip [9]. Different metals and glasses have to be investigated and selectively removed without destroying functional information of the IC [8].

Digitalizing and imaging is done in a SEM or FIB derivate in current state-of-the art reverse-engineering. With modern technologies sizes hitting the diffraction limit of optical microscopes, are more advanced visualizing tools mandatory. One the one hand this has the drawback of a moderate investment, but on the other hand can result in smaller images when the color information from optical images drop out. During the image acquiring, a brightness yield from the metals, to the vias and a brightness difference to the background is created due to different substance (electrical-)properties. A clear brightness yield from the SEM/FIB images is beneficial for the post-processing as it allows to distinguish between vias, wires and Spin-on dielectric (SOD), shown in Fig. 6.6.

---

[1]passivation, often SiO_2.

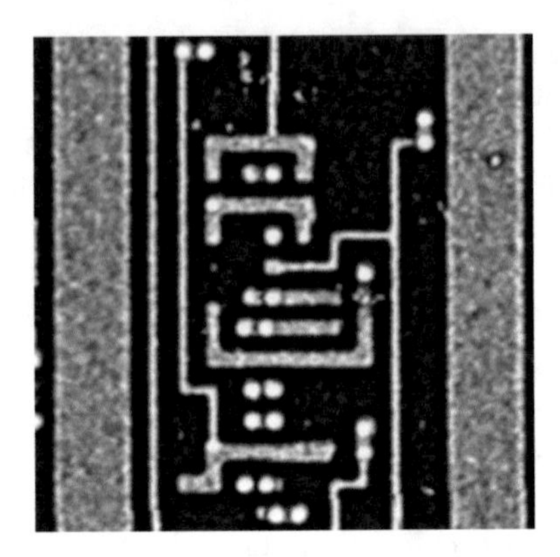

Fig. 6.6 The brightness allows to distinguish between wires, vias, and the SOD. Metal 1 in an older technology is shown. The *brighter dots* are vias between Metal1 and Metal2

Post-processing is done in software after every layer has been digitalized in tile images. The tile images are stitched, vectorized, and finally reverse-engineered to get their functional interpretation. This is a very tedious and repetitive task that can be (semi-)automated to support the reverse-engineer. Different approaches for post-processing are out of scope of this work.

Voltage Contrast: By exploiting the very nature of n-wells and p-wells, a reverse-engineer can observe a brightness yield from secondary electrons or ions using a SEM or FIB [10]. Particularly, Sugawara et al. [11] demonstrate the use of Voltage Contrast (VC) to distinguish the vias connectivity with a clear brightness yield Fig. 6.7. Notably, it is not trivial in practice to obtain meaningful results from the brightness yield, especially if the ROI is large (in the worst-case the ROI covers the whole chip). In the event, the layer images raise doubts, a reverse-engineer can enhance the doping contrast [12]. The VC shown in by Sugawara et al. [11] can be automatically included during the delayering process with a SEM, which is a state-

Fig. 6.7 Reversing stealthy dopant-level circuits. A brightness yield indicates the possible dopant regions. Taken from [11]

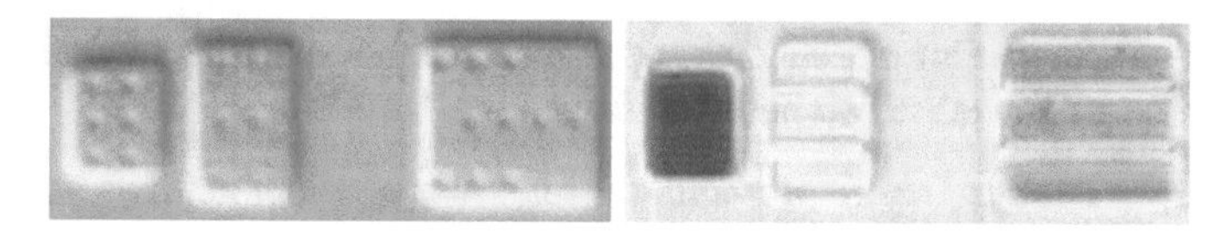

Fig. 6.8 Dash Etching. The *right* picture shows a CMOS cell with p^{+} dopant regions, stained with a *blue/green* effect depending on the applied etching time and the dopant concentration. Figure taken from [14]

of-the-art hardware reverse-engineering equipment due to the shrinking technology size.

Chemical Etching and Staining Distinguishing the dopant characteristics is often accompanied by measurement of the dopant concentration. This technique is commonly employed in failure analysis and quality control processes of silicon wafer vendors. Based on chemical etch rates or chemical staining, the dopant area can be distinguished [12–14]. For example, the chemical optimal *dash etching* exhibits different colors of p-regions and n-regions, cf. Fig. 6.8. An overview of different chemical recipes and practical approaches is given by Beck [13].

It is noteworthy that a major drawback of chemical dash etching is the optical equipment. This limits the reverse-engineer to large areas, due to the optical diffraction limit. Particularly, might become a challenge for future shrinking technology sizes, where the (stealthy) dopant areas shrink with the cell size. Advanced techniques to measure etchant rates, e.g., with a SEM should be considered.

Scanning Microscopy: In order to detect single point defects or local faults covering a few atoms of impurities, mainly two techniques for lateral doping sensitivity profiling have been established: Scanning Capacitance Microscopy (SCM) and Scanning Spreading Resistance Microscopy (SRRM). While SCM is based on capacitance differences in the substrate, SRRM is derived from the Atomic Force Microscopy (AFM) [15]. As a consequence of their small-area approaches and the required equipment, they are not recommended for identification of stealthy dopant areas. They are capable to do so, but take a lot of time. Nevertheless they are listed for the sake of completeness.

6.3 Netlist-Level Obfuscation

The successful extraction of a chip's netlist has various implications ranging from counterfeiting/cloning to technology espionage, cf. Sect. 6.1. However in several scenarios, the adversary possesses the netlist in the form of hard or soft IP cores or an untrusted foundry obtains the netlist via the chip's blueprint. To counteract IP piracy, several counterfeit avoidance methods such as secure split test and the use of Physical Unclonable Functions (PUFs) were proposed. Additionally, watermarking and

IP protection schemes are a related strand of research, however this work focuses on netlist reverse-engineering and netlist obfuscation.

Adversary Model: Before presenting the details regarding netlist-level reverse-engineering and obfuscation transformations, we briefly recap the adversary model in this scenario. We assume that the adversary has access to the flattened gate-level netlist without any a priori high-level information such as synthesis options or hierarchy structures. The high-level adversarial goal can be coarsely defined as information disclosure of how a design works in detail, in order to leverage further attacks.

6.3.1 Netlist Reverse-Engineering Techniques

In the following, we summarize the state-of-the-art in the field of algorithmic reverse-engineering of gate-level netlists. A discussion of the available techniques is vital in order to analyze the strength of an obfuscation technique. Furthermore, an overview of all published methods supports the classification of a reverse-engineering task by means of time.

In 1999, Hansen et al. [16] reported several strategies for a human reverse-engineer to extract high-level information from the ISCAS-85 benchmark suite. The strategies include the identification of common library components such as decoders or adder units and the analysis of repeated modules such as in data-path circuits. Shi et al. [17] introduced a technique to algorithmically extract Finite State Machines (FSMs) from a flattened netlist based on their inherent implementation structure. In particular, FSMs are detected based on an *enable tree* as well as *strongly connected component* identification approach. This technique is employed in the subsequent work of Shi et al. [18], where the netlist (with eliminated FSM) is analyzed and its functional modules extracted. In 2012, a technique for matching an unknown subcircuit against abstract library components was introduced by Li et al. [19]. The technique is based on pattern mining of the simulation traces as well as model checking. The subsequent work by Li et al. in [20] identified word-level structures which provides a more abstract, high-level view of the design. This work is furthermore taken as a basis for algorithmic component identification such as counters, register files and adders, etc. [21, 22]. A challenging task for functional identification is the potentially permuted input mapping of the reference circuit and the design under investigation. Gascón et al. [23] addressed this problem with a template-based approach in 2014.

After this brief recap of (semi-)algorithmic reverse-engineering capabilities, we highlight various obfuscation techniques. Furthermore, we discuss their advantages and limitations according to the published reverse-engineering techniques. First, we present control and data flow obfuscation strategies and second, reconfiguration-based methods.

6.3.2 Control Flow Obfuscation

A notable challenge from a reverse-engineer's point of view is to make sense of the design's control flow to disclose information how different modules interact with each other. Control flow obfuscation refers to a set of transformations to hamper this analysis, particularly by modification of an FSM.

Hardware Metering refers to a conglomeration of tools and security protocols to enable the design house the post-manufacturing control of a produced device. In particular, this methodology introduced in 2001 allows an unique way to identify each IC by a passive or active fingerprint [24]. Note that this strand of research is also related to obfuscation as the identification circuitry should be hard to reverse-engineer.

Internal active hardware metering can be seen as a form of control flow obfuscation [25]. The original FSM of the design is augmented by several states, cf. Fig. 6.9. In particular, the initial value of the FSM registers is determined by the output of a PUF. Only if the correct input sequence (and thus the correct traversal of the FSM states) is given, the augmented FSM ends up in the initial state of the original FSM and hence the design operates correctly. A key feature of this technique is that the number of Flip Flops (FFs) influences the number of possible states exponentially and hence provides a lightweight solution to the issue of an unique IC identifier.

A related technique is combinatorial locking (external active hardware metering). This technique extends combinatorial logic networks with the addition of XOR/XNOR nodes [26] or the gate is hidden in reconfigurable logic [27]. Only if the correct key value is applied to the input of the added nodes, the circuit is equivalent to the original one. However the claimed security of several subsequent works in this field is challenged by the recent work of Subramanyan et al. [28]. The proposed attack is based on satisfiability checking that practically unlocked the vast majority

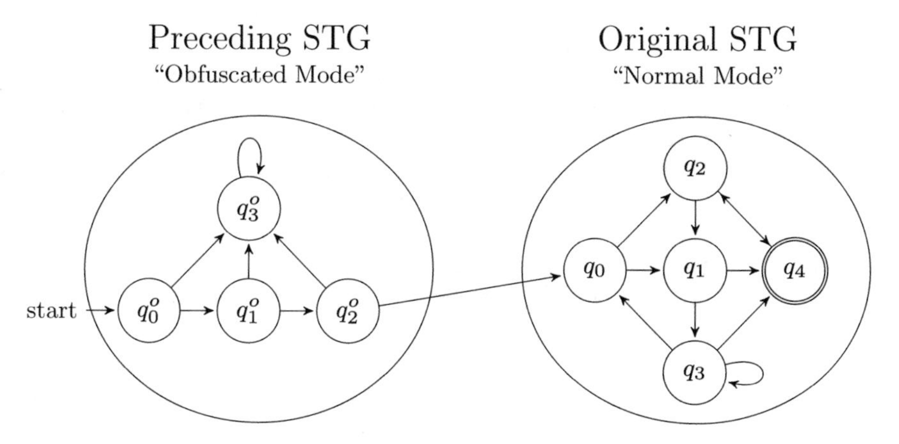

Fig. 6.9 Example: FSM Obfuscation with preceding State Transition Graph (STG) based on Fig. 1a in [32]

of allegedly locked designs. A detailed summary and discussion of the diverse hardware metering techniques is outside the scope of this work and hence the interested reader is referred to [29].

A similar FSM-based obfuscation technique to provide anti-piracy features such as authentication was proposed by Chakraborty et al. in 2008 [30]. An FSM is added to the circuitry whose inputs are the primary design inputs and it has one output. Furthermore this output is XORed with a few selected nodes of the design. Consequently, the FSM outputs a logical one as long as the correct input sequence is not applied to the primary input and only for the correct sequence the FSM transits into the state that outputs a logical zero, so that the design is equivalent to the unobfuscated one. In a subsequent work [31], the FSM output is extended by an additional signal that represents the output of a logical OR of the input variables. Later on, the method was applied to Register Transfer Level (RTL) via synthesis, application of the obfuscation on netlist-level, and subsequent decompilation to generate the obfuscated RTL [32].

Limitations: All denoted techniques have the fundamental limitation that merely the control flow (via the FSM) is obfuscated, but the inherent circuit structure for subcomponents is preserved (even if gates are appended to the output of combinatorial subcomponents). Thus, a practical evaluation is vital regarding the influence of the different automatic techniques in the presence of control flow obfuscation. Furthermore, all enumerated techniques should be evaluated regarding the FSM reverse-engineering technique by Shi et al. [17]. Particularly, all security analyses do not address the issue of reverse-engineering from the last state of the obfuscation circuitry that transits to the original initial state of the design to the best of the authors knowledge. As the design is somehow locked for an invalid input sequence, an adversary would search for conditions such as multiplexers or enable signals where a meaningful output is generated (or at least a larger set of node influences the primary output), e.g., based on established techniques such as SAT solvers. Depending on the state transition graph, an inversion may result in an exponential number of input pattern candidates, however the complexity should be evaluated practically.

Another fundamental limitation is the structure of the obfuscation circuitry itself. For example, Chakraborty et al. [31] utilize a special enable signal in their technique. First, the signal that enables the correct behavior has a large fan-out cone and its target gates are XOR elements. Second, each node in the set of selected nodes where an XOR gate is added to the output is chosen by a metric. Such inherent structures leak information regarding the implementation and might be identified using methods such as pattern matching or SAT solvers. Overall, all enumerated techniques should be evaluated regarding the statements in the limitation as well as the FSM reverse-engineering technique by Shi et al. [17].

6.3.3 Combined Data and Control Flow Obfuscation

To address the fundamental limitations of control flow obfuscation transformations, several works combined the FSM-based alteration with data flow obfuscation to generated malformed output instead of locking the device as described in the following.

In 2010 Chakraborty and Bhunia [33] presented a technique that partially consists of the prior outlined FSM alteration. Particularly, the authors demonstrated how the FSM can be interwoven with the design in order to hamper isolation of the FSM. In addition to the FSM obfuscation, the data flow is obfuscated through generation of phony output, if the system is not in a valid state (depending on the primary input sequence). This is realized by assignment of different arithmetic/logical functions to the output of the obfuscated module.

A further combined obfuscation transformation was presented by Li et al. in [34]. The key element of their methodology is the incorporation of the entire design and not only the FSM. Thus, also general circuitry such as adders or memory circuitry is transformed by the obfuscation. To be more precise, the obfuscation strategy is based on several methods that moves registers in sequential circuitry, encodes the circuit with a bijective function that is applied before and after the register stage, and addition of logic conditions under that a register value is updated. Sergeichik et al. adapted the concept of opaque predicates for hardware in [35]. The underlying principle of opaque predicates is to generate a constant output during runtime in order to hamper static analyses. Particularly, the authors insert special constant generating circuitry on the Hardware Description Language (HDL)-level, e.g., an Linear Feedback Shift Register (LFSR) where all FF values are zero or a latch-based circuit.

Limitations: Although the combination of control and data flow obfuscation definitely increases the reverse-engineer's efforts, the denoted obfuscation circuitries are static by nature. If the reverse-engineer makes sense of a structural obfuscated subcircuit, then this subcircuit will not change its functionality at some subsequent point in time. Notably, these described obfuscation techniques focus on ASICs and not on field-programmable hardware such as field-programmable gate arrays (FPGAs). Similar to the control flow obfuscation transformations, the proposed techniques were not evaluated regarding publicly known algorithmic reverse-engineering approaches, cf. Sect. 6.3.1. The decrease of information disclosure by these automatic techniques could improve the justification for the proposed obfuscation transformations.

6.3.4 Reconfiguration Obfuscation

In order to change the designs' appearance during runtime, several works exploit reconfiguration features to obfuscate a design. Notably, this methodology requires runtime field-programmable hardware features, however it addresses the generic limitation of the prior described techniques.

Porter et al. proposed an obfuscation transformation based on dynamic polymorphic reconfiguration in [36]. The underlying principle is the gate replacement implemented by the dynamic reconfiguration feature of FPGAs as well as Look-up tables (LUTs). In particular, the different gates are realized by different configurations of the LUTs. Furthermore, signals are added to the design in order to hide function signatures. To preserve the semantic of the obfuscated function, a recovery key is utilized and subsequently added to the output of the reconfigured circuit. In 2013 Gören et al. extended an FSM-based obfuscation technique (see Sect. 6.3.2) with PUFs and a dynamic reconfiguration scheme, in order to provide a low-cost FPGA bitstream protection [37]. The FSM state transition depends several PUF instances that are implemented in distinct partial configuration bitstreams reconfigured during runtime. Depending on the stored PUF outputs, the design is either locked or unlocked.

Limitations: The reconfiguration features provide a significant advantage as the adversary has to reverse-engineer has to analyze a reconfigurable part for each partial design. However, the work by Porter et al. can be simulated and thus reverse-engineered (certainly with increased efforts). Similarly to the control flow-based obfuscation, the work by Gören et al. suffers from the generic limitation that only the FSM is transformed by the obfuscation (and the rest of the design remains unchanged).

6.4 Conclusion

Hardware obfuscation techniques are demanded by the industry to hamper IP piracy and technological espionage. Particularly, obfuscation transformation aim to increase the adversary's efforts in reverse-engineering a target device or design. In this chapter, we addressed hardware obfuscation at the layout levels as well as the netlist level. In terms of layout-level obfuscation, relatively few public information is available. While obfuscation is been in use for many years, how exactly—and how effectively—it is being used is not discussed publicly. Only very recently was the first scientific paper published in that regard. Most of these layout-level obfuscation techniques are based on the idea to construct camouflage gates based on changes of the dopant polarity in the active area. However, several visual reverse-engineering techniques exists that can detect the dopant polarity in an active area. These techniques require additional steps and equipment compared to traditional layout reverse-engineering and hence can make reverse-engineering considerably harder. However, none of the layout level obfuscation techniques can completely prevent reverse-engineering. In general, form a research perspective, many unanswered questions remain in this area. Furthermore, the fact that the companies that specialize on reverse-engineering do not reveal their techniques to the public, estimating the cost of reverse-engineering a design with and without layout obfuscation is currently very difficult.

Several works proposed diverse methods to realize obfuscation transformations on the netlist-level ranging from control flow-based techniques to reconfiguration-based methods. However, we identified various limitations for the different approaches especially regarding the security considerations. The coarse adversary model for obfuscation should be regarded in detail with respect to the system model and the defensive goal. Particularly, reverse-engineering of a design in order to disclose IP and patching of a design in order to eliminate locking features are elementary different goals. Furthermore, a fundamental issue for the majority of the analyzed works is the omission of the automatic reverse-engineering techniques, cf. Sect. 6.3.1. Particularly, an evaluation of the diverse obfuscation transformations combined with the reverse-engineering techniques is viable for future research in this area. Additional to the obfuscation transformations, the reverse-engineering techniques have to be further explored in order to improve both the obfuscation transformations as well as the modeling of real-world adversarial capabilities.

Overall, hardware obfuscation provides a powerful set of tools to increase an adversary's reverse-engineering efforts. To really understand the level of obfuscation and security achieved by the different techniques, it is also crucial to understand the capabilities of reverse-engineers. Unfortunately, often the public knowledge of the state-of-the-art reverse-engineering techniques is limited since reverse-engineering companies do not publish their methods. In many cases, the real advantage of the different obfuscation technologies are therefore hard to estimate in practice. In general, there are still more open then solved research questions in the area of hardware obfuscation.

References

1. H. Pechar, Circuit to prevent pirating of an mos circuit, US Patent 4,583,011, 15 April 1986
2. R.P. Cocchi, J.P. Baukus, L.W. Chow, B.J. Wang, Circuit camouflage integration for hardware ip protection, in *Proceedings of the 51st Annual Design Automation Conference (DAC 14)*, New York, NY, USA, 2014 (ACM, 2014), pp. 153:1–153:5
3. S. Malik, G.T. Becker, C. Paar, W.P. Burleson, Development of a layout-level hardware obfuscation tool, in *VLSI (ISVLSI), 2015 IEEE Computer Society Annual Symposium on July 2015*, pp. 204–209
4. J. Rajendran, M. Sam, O. Sinanoglu, R. Karri, Security analysis of integrated circuit camouflaging, in *Proceedings of the 2013 ACM SIGSAC Conference on Computer and Communications Security* (ACM, 2013), pp. 709–720
5. M. Shiozaki, R. Hori, T. Fujino, Diffusion programmable device: the device to prevent reverse engineering. IACR Cryptology ePrint Archive **2014**, 109 (2014)
6. G.T. Becker, F. Regazzoni, C. Paar, W.P. Burleson, Stealthy dopant-level hardware Trojans, in *Cryptographic Hardware and Embedded Systems (CHES 2013)* (LNCS, Springer, 2013)
7. A. Bogdanov, L.R. Knudsen, G. Leander, C. Paar, A. Poschmann, M.J. Robshaw, Y. Seurin, C. Vikkelsoe, Present: an ultra-lightweight block cipher, in *Proceedings of the 9th International Workshop on Cryptographic Hardware and Embedded Systems (CHES 07)* (Springer, Berlin, Heidelberg, 2007), pp. 450–466

8. S.E. Quadir, J. Chen, D. Forte, N. Asadizanjani, S. Shahbazmohamadi, L. Wang, J. Chandy, M. Tehranipoor, A survey on chip to system reverse engineering. ACM J. Emerg. Technol. Comput. Syst. (JETC), **13**(1), 1–34 (2016). Article 6
9. C. Kison, J. Frinken, C. Paar, *Cryptographic Hardware and Embedded Systems—CHES 2015: 17th International Workshop*, Saint-Malo, France, 13–16 Sept 2015, Proceedings, chapter Finding the AES Bits in the Haystack: Reverse Engineering and SCA Using Voltage Contrast (Springer, Berlin, Heidelberg, 2015), pp. 641–660
10. S. Prejean, B. Davis, L. Herlinger, R. Johnson, R. Parente, M. Santana, *Special Techniques for Backside Deprocessing* (Desk Reference. A S M International, In Microelectronics Failure Analysis, 2011)
11. T. Sugawara, D. Suzuki, R. Fujii, S. Tawa, R. Hori, M. Shiozaki, T. Fujino, Reversing stealthy dopant-level circuits, in *Cryptographic Hardware and Embedded Systems (CHES 2014), LNCS*, vol. 8731 (Springer, 2014), pp. 112–126
12. E. Le Roy, R. Pajak, F. Baiocchi, Dopant imaging on front surface of silicon devices with a coaxial photon-ion column, in *ISTFA 2005: Proceedings of the 31st International Symposium for Testing and Failure Analysis* (ASM International, 2005)
13. F. Beck, *Integrated Circuit Failure Analysis: A Guide to Preparation Techniques* (Wiley-Interscience, 1998)
14. Silicon PrOn: silicon just the way you like it. Last visited Feb 2016. http://siliconpr0n.org/wiki/doku.php?id=start
15. D.K. Schroder, *Semiconductor Material and Device Characterization* (Wiley-Interscience, 2006)
16. M.C. Hansen, H. Yalcin, J.P. Hayes, Unveiling the ISCAS-85 benchmarks: a case study in reverse engineering. IEEE Des. Test Comput. **16**(3), 72–80 (1999)
17. Y. Shi, C.W. Ting, B.-H. Gwee, Y. Ren, A highly efficient method for extracting fsms from flattened gate-level netlist, in *International Symposium on Circuits and Systems (ISCAS 2010)*, Paris, France, 30 May–2 June 2010, pp. 2610–2613
18. Y. Shi, B.H. Gwee, Y. Ren, T. Khaing, C.W. Ting, Extracting functional modules from flattened gate-level netlist, in *2012 International Symposium on Communications and Information Technologies (ISCIT)*, (2012) pp. 538–543
19. W. Li, Z. Wasson, S.A. Seshia, Reverse engineering circuits using behavioral pattern mining, in *2012 IEEE International Symposium on Hardware-Oriented Security and Trust, HOST 2012*, San Francisco, CA, USA, 3–4 June 2012, pp. 83–88
20. W. Li, A. Gascón, P. Subramanyan, W.Y. Tan, A. Tiwari, S. Malik, N. Shankar, S.A. Seshia, Wordrev: finding word-level structures in a sea of bit-level gates, in *2013 IEEE International Symposium on Hardware-Oriented Security and Trust, HOST 2013*, Austin, TX, USA, 2–3 June 2013, pp. 67–74
21. P. Subramanyan, N. Tsiskaridze, W. Li, A. Gascón, W.Y. Tan, A. Tiwari, N. Shankar, S.A. Seshia, S. Malik, Reverse engineering digital circuits using structural and functional analyses. IEEE Trans. Emerg. Top. Comput. **2**(1), 63–80 (2014)
22. P. Subramanyan, N. Tsiskaridze, K. Pasricha, D. Reisman, A. Susnea, S. Malik, Reverse engineering digital circuits using functional analysis, in *Design, Automation and Test in Europe, DATE 13*, Grenoble, France, 18–22 Mar 2013, pp. 1277–1280
23. A. Gascón, P. Subramanyan, B. Dutertre, A. Tiwari, D. Jovanovic, S. Malik, Template-based circuit understanding, in *Formal Methods in Computer-Aided Design, FMCAD 2014*, Lausanne, Switzerland, 21–24 Oct 2014, pp. 83–90
24. F. Koushanfar, G. Qu, Hardware metering, in *Proceedings of the 38th Design Automation Conference, DAC 2001*, Las Vegas, NV, USA, 18–22 June 2001, pp. 490–493
25. Y. Alkabani, F. Koushanfar, Active hardware metering for intellectual property protection and security, in *Proceedings of the 16th USENIX Security Symposium*, Boston, MA, USA, 6–10 Aug 2007
26. J.A. Roy, F. Koushanfar, I.L. Markov, EPIC: ending piracy of integrated circuits, in *Design, Automation and Test in Europe, DATE 2008*, Munich, Germany, 10–14 Mar 2008, pp. 1069–1074

27. A. Baumgarten, A. Tyagi, J. Zambreno, Preventing IC piracy using reconfigurable logic barriers. IEEE Des. Test Comput. **27**(1), 66–75 (2010)
28. P. Subramanyan, S. Ray, S. Malik, Evaluating the security of logic encryption algorithms, in *IEEE International Symposium on Hardware Oriented Security and Trust, HOST 2015*, Washington, DC, USA, 5–7 May 2015, pp. 137–143
29. F. Koushanfar, *Introduction to Hardware Security and Trust*, chapter Hardware Metering: A Survey (Springer, New York, NY, 2012), pp. 103–122
30. R.S. Chakraborty, S. Bhunia, Hardware protection and authentication through netlist level obfuscation, in *2008 International Conference on Computer-Aided Design, ICCAD 2008*, San Jose, CA, USA, 10–13 Nov 2008, pp. 674–677
31. R.S. Chakraborty, S. Bhunia, HARPOON: an obfuscation-based soc design methodology for hardware protection. IEEE Trans. CAD Integr. Circuits Syst
32. R.S. Chakraborty, S. Bhunia, Security through obscurity: an approach for protecting register transfer level hardware IP, in *IEEE International Workshop on Hardware-Oriented Security and Trust, HOST 2009*, San Francisco, CA, USA, 27 July 2009, pp. 96–99
33. R.S. Chakraborty, S. Bhunia, RTL hardware IP protection using key-based control and data flow obfuscation, in *VLSI Design 2010: 23rd International Conference on VLSI Design, 9th International Conference on Embedded Systems*, Bangalore, India, 3–7 Jan 2010, pp. 405–410
34. L. Li, H. Zhou, Structural transformation for best-possible obfuscation of sequential circuits, in *2013 IEEE International Symposium on Hardware-Oriented Security and Trust, HOST 2013*, Austin, TX, USA, 2–3 June 2013, pp. 55–60
35. V. Sergeichik, A. Ivaniuk, Implementation of opaque predicates for fpga designs hardware obfuscation. J. Inf. Control Manage. Syst. **12**(2) (2014)
36. R. Porter, S.J. Stone, Y.C. Kim, J.T. McDonald, L.V.A. Starman, Dynamic polymorphic reconfiguration for anti-tamper circuits, in *19th International Conference on Field Programmable Logic and Applications, FPL 2009*, Prague, Czech Republic, 31 Aug–2 Sept 2009, pp. 493–497
37. S. Gören, O. Ozkurt, A. Yildiz, H.F. Ugurdag, R.S. Chakraborty, D. Mukhopadhyay, Partial bitstream protection for low-cost fpgas with physical unclonable function, obfuscation, and dynamic partial self reconfiguration. Comput. Electr. Eng. **39**(2), 386–397 (2013)

Chapter 7
An Application of Partial Hardware Reverse Engineering for the Detection of Hardware Trojan

Franck Courbon

7.1 Introduction

In Hardware Trojan (HT) detection research activities, several families of techniques have been studied and proposed [1]. The choice of the detection technique can be chosen based on performing a vulnerability analysis regarding HT insertion risk within the entire production chain. Once the associated chain of trust is known, possible locations of interest for applying the detection are identified. The goals of the detection technique can vary depending on the final application of the circuit.

In this study, the case of a secure circuit development supply chain such as that used for smart card production, is taken as an example. The role of a software developer who receives the circuit from a chip maker is considered as a hypothetical scenario. The selection of the HT detection technique is driven by the following parameters: detection time available, type of HT detected, detection success rate, size of the HT detected, robustness regarding process variation, golden model availability, and the cost of detection. They have different weights depending on the given IC application.

The main contribution of our work is to introduce a novel HT detection methodology based on visual inspection of a single layer of Integrated Circuits (ICs). This provides a global image of the circuit where only the transistors' active regions are visible. Every standard cell (without knowing its function) is localized and then compared with a circuit reference. We offer a low-cost, fast and efficient partial reverse engineering methodology combined with an image processing based final detection.

F. Courbon (✉)
Computer Laboratory, University of Cambridge, Cambridge, UK
e-mail: frc26@cam.ac.uk

F. Courbon
Gemalto Security Evaluation Labs, La Ciotat, France

F. Courbon
Ecole des Mines de Saint-Etienne, CMP-GC/LSAS, Gardanne, France

L. Bossuet and L. Torres (eds.), *Foundations of Hardware IP Protection*,
DOI 10.1007/978-3-319-50380-6_7

The method is applied to a real case of detection on university test chips: a genuine circuit (claimed to be manufactured under a fully trusted environment) and a Hardware Trojan (HT) infected chip. Both circuits are chemically prepared to access the transistors' active region before reconstructing the entire surface of this single chip layer. It results in an image so-called in this chapter the 'entire IC transistors' active region image'. Image processing allows then a comparison between the entire IC transistors' active region image and the trusted circuit reference image. On top of detecting the HT, the number of modified gates and their localization are also retrieved. This application highlights the efficiency of our methodology while being in line with the integrated circuit's life cycle constraints.

7.2 Integrated Circuits, Malicious Hardware Modifications and Base of Retro-Engineering

7.2.1 Smart Card Like Integrated Circuits

As Hardware Trojans are physically implemented in integrated circuits, we first describe their structure, taking as an example a 90 nm technology node current smart card type IC. Several elements from the transistor to a specific function are manufactured using different physical layers from the active regions to the top metal layer, Fig. 7.1. The substrate, made of Silicon (Si.), allows manufacturing of the different lithography steps and handling of the component. Figure 7.1 highlights the weak thickness of the different physical layers expressed in µm whilst the width and the

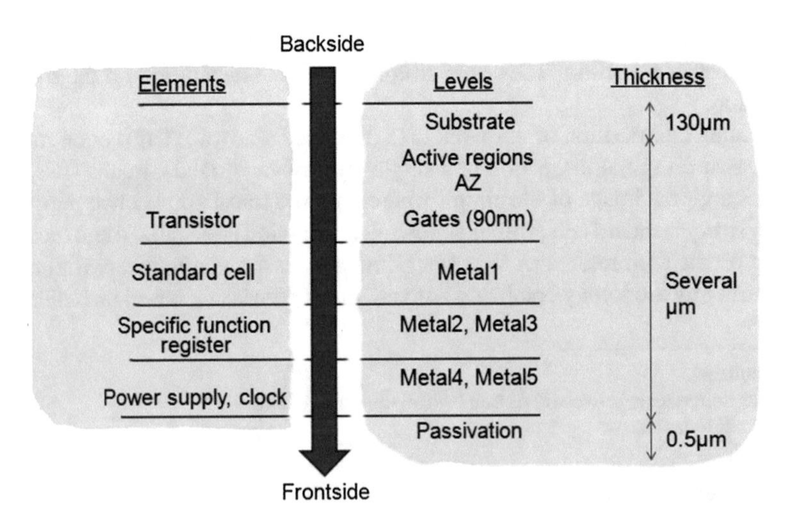

Fig. 7.1 Typical smart card like IC elements and layers

length of a device are expressed in several mm, from chip backside to chip frontside we find:

- A 130 μm Silicon substrate thickness (smart card), covers the main volume of the circuit, and has no active properties,
- a transistor level made of active regions (wells and doped area) for transistor's drain and source and poly-Si for transistor's gate,
- a standard cell level made of transistors and metal1 layer,
- a specific function or register level made of standard cells interconnected in metal2 and metal3,
- power and clock signals are generally present in metal4 and metal5,
- insulators are present between each layer (a via interconnects two successive layers).

7.2.2 Reverse Engineering

Reverse engineering techniques have historically been developed to perform the opposite of a typical process flow used to build integrated circuits (IC). Nowadays, IC reverse engineering can be used for different purposes. A given company can analyze its own product for validation, debugging, Hardware Trojan detection, or failure analysis. The same company might also be interested in reverse engineering competitors' products for other needs, such as looking for IP infringements, extracting obsolete circuit specifications, or reverse costing. Reverse engineering can be applied from a complete product or system composed of a printed circuit board with multiple integrated circuits to a single transistor parameter, as shown in Fig. 7.2.

Typical IC reverse engineering methods need a collection of images coming from each metal layer—up to around 15 for advanced semiconductor processes—that have to be individually and successively prepared. This requires a high-level skill, years of expertise, expensive equipment, high precision, and time. Once these layers' images are acquired, the main remaining task is to reconstruct the connections between each layer and extract the chip's netlist (description of the connectivity of an integrated circuit). Some papers deal specifically with image processing to recover integrated

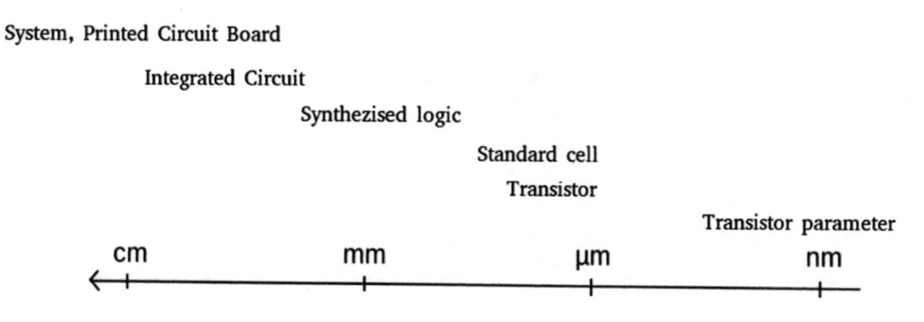

Fig. 7.2 Reverse engineering range of resolution

circuit information [2–5]. Some specialized companies achieve automated reverse engineering by using data extraction tools not readily available to the (academic) research community. The cost of a full chip reverse engineering is estimated to be several tens of thousands euros for a 130 nm technologic node chip containing 100 k logic gates. The global cost can also be evaluated in terms of processing time and delay, the need to send samples out of premises, the need for experts, high-end tools, and the need to have multiple samples.

7.2.3 Hardware Trojan Taxonomy and Threats

Hardware Trojans have various properties [6], as shown in Table 7.1, and no detection technique covering all kind of HTs has been proposed so far. For instance, detection methods such as Side Channel Analysis (SCA) and delay-based detection, affected by process and measure variations, reach a low detection rate, require large data sets and may not cover any HTs size or activity. Closer to our approach, only simulations have been done on a partial reverse engineering hardware trojan detection [7] and only top metal layer hardware trojan detection tests have been performed on FPGA design tool images [8].

One of the key factors defining a HT detection method is the detection rate. As Hardware Trojans can have various physical characteristics, even a 'full' hardware reverse engineering which is cost and time expensive [9] will not be '100 %' efficient. For instance, it does not cover parametric Hardware Trojans [10]. Inherently being a solution for HT detection, we choose to investigate and develop partial hardware reverse engineering capabilities. The goal is to visualize, save, and analyze physical elements of ICs that are usually non accessible by different microscopy means.

Table 7.1 Hardware Trojan taxonomy

Design phase	Abstraction level	Activitation	Effects	Location	Physical characteristic
Specification	System level	Always on	Change function	Processor	Distribution
Design	Development environment	Internally time based triggered	Degrade perf.	Memory	Size
Fabrication	RT level	Internally physical cond. triggered	Leak information	I/O	Parametric type
Test	Gate level	Externally user triggered	Denial-of-service	Power supply	Functional type
Assembly and Package	Transistor level	Externally component triggered		Clock	Layout-same structure
	Physical level				Layout-change structure

7.3 Accessing Information by Microscopy Means

7.3.1 Bacskide Non-destructive Imaging

It has to be noted that state of the art backside non-destructive imaging, Fig. 7.3, does not achieve a resolution good enough to distinguish standard cell modification. Through this document, we will see a choice of a frontside destructive approach which is fast, low-cost and efficient for our need.

7.3.2 Layer of Interest to Be Accessed

Also along this document, we will see that the part of interest (the transistors' active layer) is located between the metal layers and the Silicon substrate. Using visible light microscopy, Fig. 7.4, does not permit to reach the layer of interest.

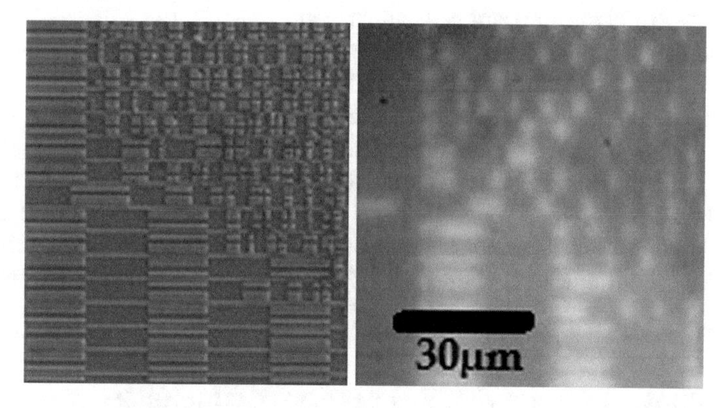

Fig. 7.3 No distinguishable differences within standard cells using backside imaging

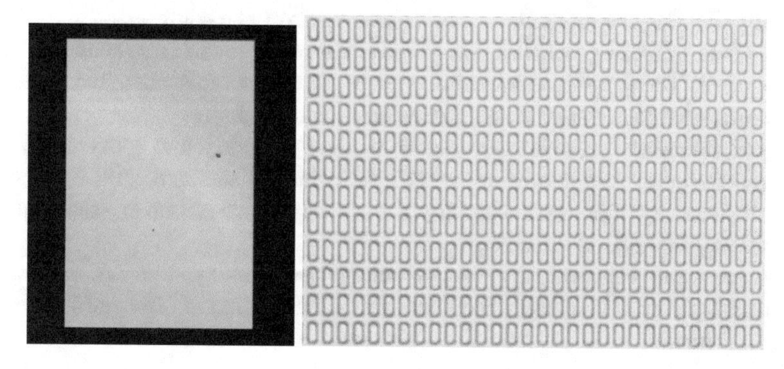

Fig. 7.4 Frontside and backside visible light acquisitions of an integrated circuit

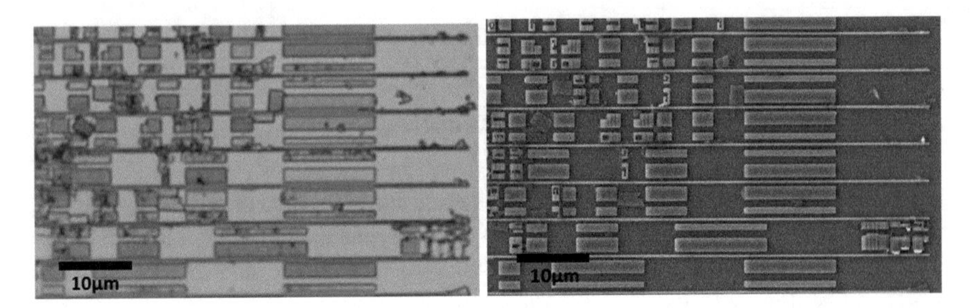

Fig. 7.5 Optical and SEM comparison

7.3.3 Choosing Electron Microscopy

Nowadays, a large number of imaging possibilities exist. Our choice directly goes for the Scanning Electron Microscope as its intrinsic features contribute to the success of the methodology. Nevertheless, it is noted that optical microscopes can also give enough information to distinguish modified and non-modified cells as seen in Fig. 7.5, where 130 nm standard cells are present.

SEM and optical images are obtained in seconds and acquisition routine can be defined without any operator intervention. Nevertheless, SEM permits a large depth of view, no nonuniform illumination matter. It has to be noted that SEM is, at the moment, seen as bespoke equipments in some security schemes [11] despite their everyday use in fields like materials or natural sciences. As a consequence, it is possible to rent such equipment in various mutualized platforms or universities for less than hundreds euros per hour. Finally, processing all those individual images is achievable without much effort in terms of competences, resources, cost, and time.

7.4 Proposal of a Novel HT Detection Methodology

After highlighting the integrated circuit's structure, reverse engineering capabilities and Hardware Trojan properties, we propose a novel approach for detecting Hardware Trojans. The proposed methodology is based on previously achieved works in the security characterization of integrated circuits field [12].

A partial reverse engineering approach has been developed to spot critical cells in terms of security and permitting to retrieve 'prints' of standard cells, Fig. 7.6. To check the integrity of integrated circuits, the same approach is used and the complete methodology is made of three steps.

The proposed Hardware Trojans detection methodology, Fig. 7.6, consists in preparing the circuit (1st step), acquiring and aligning images (2nd step) and comparing information retrieved from the device under test with a genuine reference (3rd step).

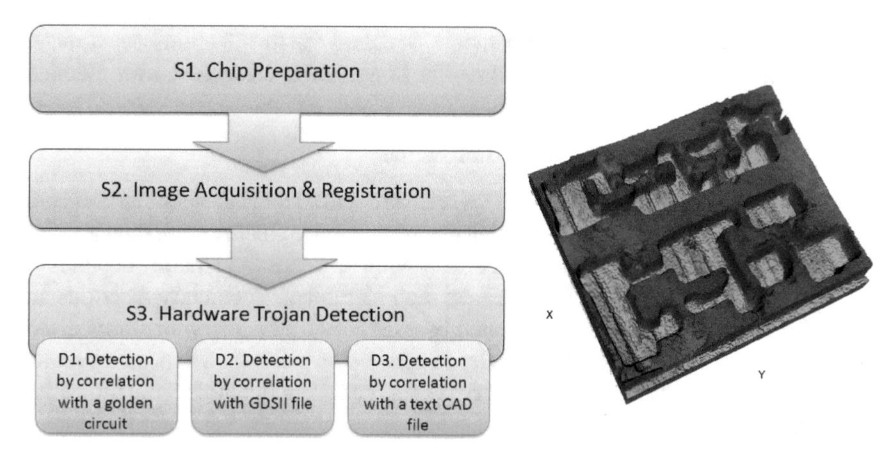

Fig. 7.6 Three steps methodology proposal based on the extraction of standard cell 'prints' [13]

7.4.1 Step 1: Sample Preparation

The chip is first prepared to access a layer where any standard cell modification or addition would be visible.

Reaching top metal layers Depending on the package type, wet etching (hot fuming nitric acid), decapsulation dedicated machine, or even sharp tools, can be used to reach the top metal layer. At this stage, it is not possible to observe the standard cells/transistors with frontside imaging: the different metal layers will obstruct the frontside acquisition, and a metal shield can also be present.

Removing all metal layers The integrated circuits thus need a preparation [14] to distinguish physical disparities in the component. We access the BULK layer using Hydrofluoric (HF) acid. The acid bath removes all di-electric present under those metal layers. ICs are simply soaked into this acid bath for several minutes. One can see the remaining Tungsten plugs lying over the remaining surface of the component. They normally form a contact between Metal1 and diffusion.

Rendering the surface free of artifacts Once the acid bath performed and the sample rinsed with acetone, it is necessary to clean the integrated circuit. To remove wet etching residues, the integrated circuit can be placed into an ultrasonic cleaner for about 5 min. The transistor's active region is now clearly distinguishable.

7.4.2 Step 2: Automatic Image Alignment and Registration

The image acquisition is performed with a Scanning Electron Microscope (SEM). Images are individually saved and then registered together to obtain a global image;

the entire IC transistors' active region image. Standard cell localizations are extracted from the entire IC transistors' active region image to be compared with identical information from a genuine reference.

Setting up the Scanning Electron Microscope

Scanning Electron Microscope (SEM) properties Within a SEM chamber, a current going through a filament allows creating electrons. Those electrons are then accelerated before hitting the specimen surface. The depth of penetration mainly depends on the accelerating voltage and the specimen atomic number. Secondary electrons are emitted from the specimen and collected by a dedicated secondary electrons detector (in-lens or not). A grayscale intensity image is at last obtained depending on the number of collected electrons. Current SEM features are about 500kX magnification and about a few *nm* lateral resolution.

SEM parameters Multiple standard SEM parameters have to be considered: the type of detector, working distance, accelerating voltage, detector current, contrast, luminosity, astigmatism, focus, and scanning speed. Moreover, the user needs to set the tilt, the rotation, and the XY placement.

Multiple image acquisition

Defining area and magnification to scan Despite roughly knowing the Hardware Trojan area over our test samples, we want to gather images of the entire IC to validate our methodology. Thus, the full IC is 'framed' to be the area to acquire. The scanning area being defined, the software returns the real coordinates (in μm) under a calibrated equipment. Afterwards, the magnification is set up: the larger the magnification, the better the resolution. However, this has a direct impact on the speed of acquisition. For our purposes, the magnification has been chosen depending on the ability to distinguish cells while not being too time consuming.

Automating the acquisition Once the magnification, the image overlay and the scanning area defined, acquisition routines, including SEM actions and setup, can be written. Afterwards, those acquisitions need to be stitched together to form the entire IC transistors' active region image. The automatic image acquisition is based on a script that includes four basic principles. It defines the matrix to go through, move the stage to each matrix cell, add idle state to collect the full image information and save each displayed image.

All chip registration Images have been acquired and saved in a working directory where information about scan direction and acquisition number is saved. They are then offline registered to construct a single image of the full circuit for each case. Using open source libraries or softwares, a phase transformation based algorithms implementation answers this need.

7.4.3 Step 3: Hardware Trojan Detection

The genuine reference can either be a golden circuit or a design file. A circuit with a manufacturing flow that is fully done in-house, or more probably the first batch of a product, can be considered genuine. They do not allow an adversary to introduce a Hardware Trojan. Thus, to detect HT insertion using a genuine circuit, substep D1 in Fig. 7.6, an entire IC transistors' active region image is obtained over both circuits. Genuine and tested circuit images are then processed to identify hardware malicious modifications.

Comparing the entire IC transistors' active region image with a design file is also possible. It involves the application of pattern recognition algorithms. The technique requires either a graphical file, such as the GDSII, or a text file, such as a DEF file. The first technique consists of modifying the GDSII polygon view to get closer to the real physical shapes (or inversely obtaining polygons from the real physical shapes). The second technique comparing the entire IC transistors' active region image with a text design file (such as a DEF file) first requires a count of the occurrences of each standard cell in the circuit. The final detection compares the position of recognized standard cells (shapes) with the position included in the DEF file.

Only transistors' active regions remain after our preparation technique and are acquired over the full integrated circuit. The standard method used to retrieve logic gate functionality is to overlay the active region of the transistors, the polysilicon layer, and the first metal layer to draw the circuit schematic and then simulate it. We don't go through this typical reverse engineering process but instead we correlate the entire IC transistors' active region image and a genuine reference. More information is given in Sect. 7.6.

Image processing either permits to correlate two circuits prepared and acquired the same way or correlate data from a prepared circuit and a genuine reference. Each shape recognized in the entire IC transistors' active region image allows adding an entry to a database file. It contains the pattern number, the occurrences number and the [X, Y] coordinates for each shape recognized.

Plus, we get enough information to make a classification between cells and to optionally form an hypothesis on a gate's function. For instance, regarding other gates in the design, flip-flop gates contain one of the most important number of transistors [15]. We isolate in Fig. 7.7 a single pattern visible on the entire IC transistors' active region image but acquired at a larger magnification. The PMOS transistors' location is the side of the column having the larger width. The opposite side is NMOS transistors' location. The different transistor gates' positions after etching can be noticed. For instance, in this pattern, we are able to distinguish around 13 different gates for each MOS side, leading to a standard cell with something like 26 transistors. Therefore, this important number of transistors is representative of a flip-flop.

Fig. 7.7 Identification of a 130 nm technology node flip-flop pattern at 18k× magnification [16]

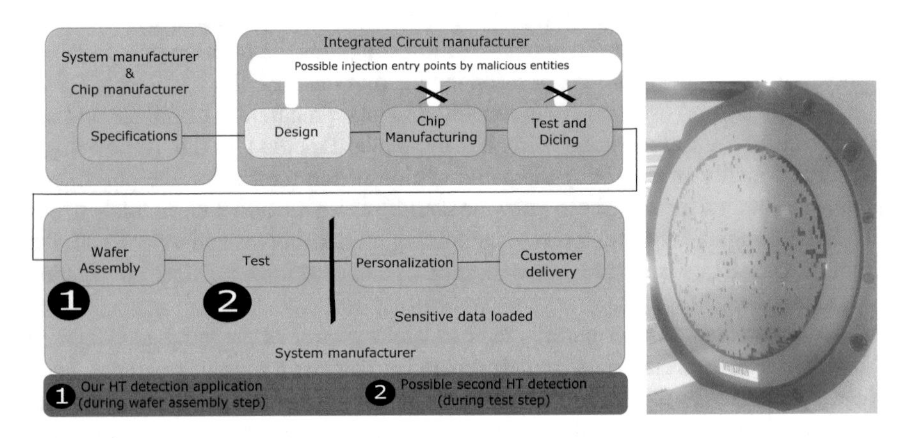

Fig. 7.8 *Left* Adapted IC flow with HT detection, *right* applied at wafer reception

7.5 Methodology Advantages

7.5.1 IC Flow Compliancy

When it comes to choosing a potential Hardware Trojan detection method, we have to consider the IC's life cycle. For instance, previous investigations [6] give the level of trust of each integrated circuit fabrication step from the specification to the IC's deployment. All steps between the IC specifications and the package test are described as untrusted. Some of those IC life cycle steps are done in-house for a system manufacturer and therefore appear trusted with our system manufacturer point of view. We adapt the supply chain with our HT detection technique proposal.

Table 7.2 Summary of method advantages

Time slot present at wafer reception
Detection only requires few hours
No yield impact
Low application cost if equipment rented
Robust versus process variations
Robust versus setup measure variations
Validated proof of concept
Application by nonexpert
Independent of the technology node

The technique is IC life cycle 'friendly' and placed at the wafer reception, as illustrated in Fig. 7.8. At last, the two steps 'Chip manufacturing' and 'Test and dicing' are those which become trusted with the application of our HT detection proposal.

7.5.2 *Industrial Advantages*

As the detection is directly based on the hardware level of the component, maximum efficiency is reachable. This efficiency rate is robust regarding both process and setup measure variations. One interesting point is that even a non-multidisciplinary expert can apply such a methodology. The application cost itself is low regarding the entire integrated circuit design flow. We also note that the fabrication yield is not impacted as only nonfunctional ICs are picked from the wafer up. The idea is to take nonfunctional ICs that are representative of the remaining functional integrated circuits. The percentage of functional ICs being checked depends on the ICs made with exactly the same flow (lithography masks and parameters, number of circuit manufactured at once (stepper), localization of nonfunctional integrated circuits). We sum up the main advantages of our HT detection methodology in Table 7.2.

7.6 The Three Different Detection Scenarios

In this Sect. 7.6, a four standard cell Hardware Trojan is simulated in order to illustrate the different detection scenarios previously reported in Fig. 7.6. The Hardware Trojan is simulated by directly adding four gates instances taken from the entire IC transistors' active region image of a 1 mm^2 hardware ciphering test circuit [17]. In this example, the entire IC transistors' active region image is obtained from 64 individual images acquired at 2.2k$\times$ magnification including a 10 % overlap.

7.6.1 Golden Circuit

Once transistors' active regions acquired, the intention to compare a couple of circuits, a genuine circuit and a circuit to test is trivial. Thus, sample preparation and image acquisition steps are performed over two different circuits: a golden circuit and a circuit to test. Mainly due to the preparation type, acquired images under the microscope can show some local differences, such as the presence of remaining superior layers at some locations. Moreover, when preparing the golden circuit and each new selected chip poses some challenges. The application of acids has to be identical in terms of time and concentration. Moreover, using the same SEM parameters may not result in the same resulting image. We report strong image intensity differences between two successive acquisitions of the same sample in Fig. 7.9.

A direct image subtraction can not check if any circuit modification has been inserted. Indeed, images are not acquired using exactly the same SEM parameters, the chip does not have the same orientation, and the scanned area is not exactly the same. Therefore, some image processing has to be applied in order to register an image according to other one and then reveal differences between the images. We highlight the four simulated additional gates in Fig. 7.10 using the golden circuit approach.

7.6.2 GDSII File

A golden model circuit may not be available, e.g., if no fully trusted manufacturing process is available. Starting from the same image, we describe how to compare the entire IC transistors' active region image with a graphical Computer Aided Design (CAD) file, the GDSII. Figure 7.11 is composed of a flip-flop standard cell GDSII where only n-wells and p-wells implanted zone is kept under the CAD tool. On the

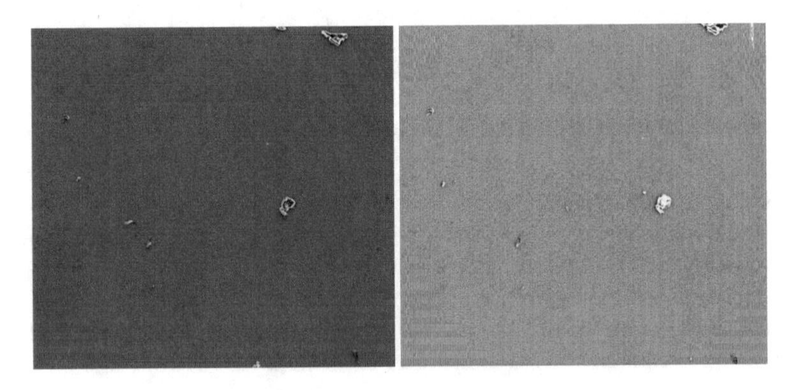

Fig. 7.9 Same chip, different SEM grayscale intensity values

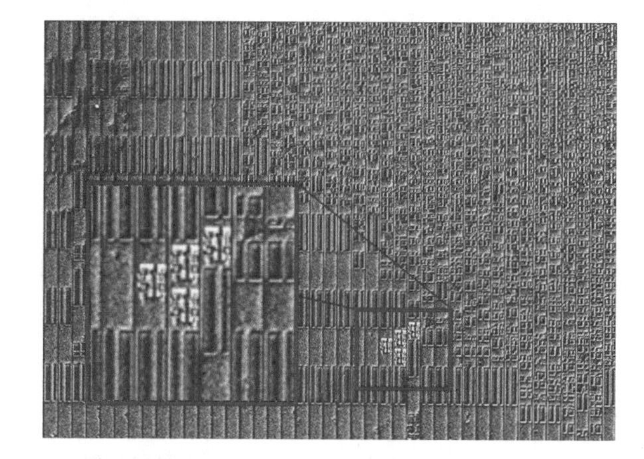

Fig. 7.10 Detecting simulated Hardware Trojans with golden circuit approach

Table 7.3 Detection case using modified GDSII data

	Localization 1	Localization 2
Physically extracted		
Modified GDSII		
NCC coefficient	Large	Small
Hardware Trojan presence	No	Yes

second shape, standard closing morphology image processing and dilation are used. It permits to facilitate the correlation to be performed over the manufactured device.

At a specific location, if the correlation coefficient significantly drops, a circuit modification is detected. The HT is detected and its location known as shown in Table 7.3.

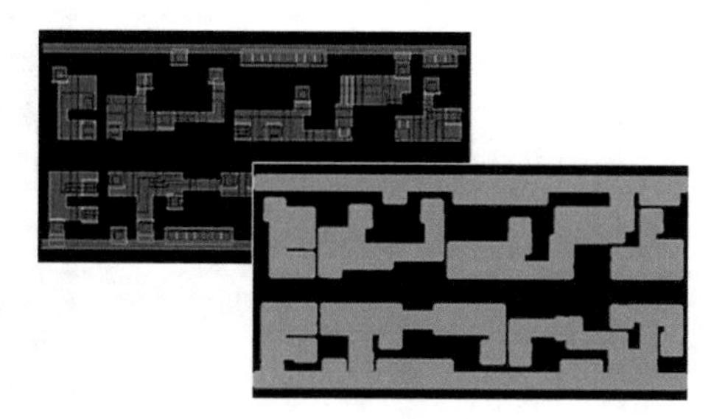

Fig. 7.11 GDSII layout and modified GDSII

Table 7.4 Detection with a text CAD file

Name	Localization in DEF file in μm	Total number of occurrences	Extracted view	Number of recognized occurrences	Localization of recognized instances in *pix*.
FF1	DX_1, DY_1, DX_2, DY_2, DX_n, DY_n	N_{rec}		N_{rec}	RX_1, RY_1, RX_2, RY_2, RX_n, RY_n

7.6.3 DEF File

Another Computer Aided Design (CAD) tool output file can be used as a reference, the DEF file. It may be easier to transmit the DEF file as this CAD tool extracted text file does not contain connection information between standard cells. In this type of file, each standard cell name is written as does its location in μm relative to the origin. The idea is to first class all the instances by standard cell types and add each similar one to obtain a number of similar shapes to retrieve over the circuit under test. For example, we obtain localizations of the given flip-flop over the complete circuit by taking a flip-flop cell instance as pattern to recognize, Fig. 7.12. A standard correlation tool based on normalized cross correlation is used [18].

Each detected pattern is visualized as a white rectangle. The pattern detection matrix is also stored into a file storing recognized elements and their [X, Y] positions.

As illustrated in Table 7.4, if the number of gates present in the DEF file, N_{occ}, differs from the number of recognized occurrences, N_{rec}, a circuit modification has been introduced within the test device. Localizations of gates should also correlate if the

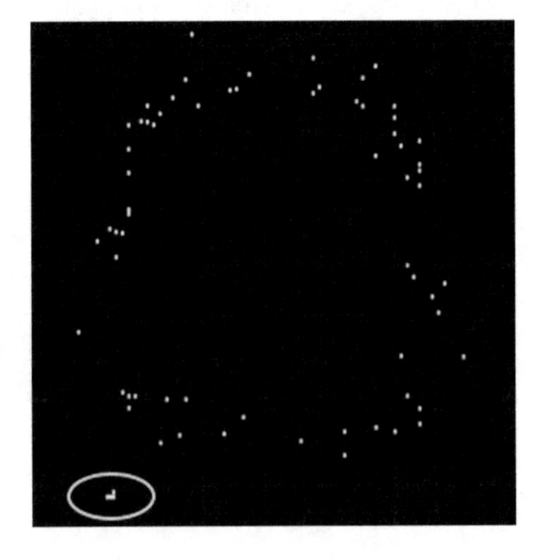

Fig. 7.12 Recognized flip-flop localizations

chip is Trojan-free. In this physically simulated HT insertion, N_{occ} is four instances smaller than N_{rec}, it corresponds to the group of four gates highlighted in Fig. 7.12.

7.7 Applying the Methodology to a Real Detection Case

Having the opportunity to work with two versions of a circuit [19], one infected and the other genuine, the application of our methodology is straightforward, as illustrated in Fig. 7.13. The two entire IC transistors' active region image has been obtained by applying the partial reverse engineering methodology.

7.7.1 *The ASICs and Equipments Used*

The circuits used to validate our HT detection methodology are hardware implementations of ciphering algorithms and hash functions. Their size is thus smaller than typical Integrated Circuits. These ICs, manufactured in a 0.18μm technology node, are 1 mm long by 1 mm wide, Fig. 7.14.

For the application of the developed methodology, HF acid and an ultrasonic cleaner permit to access transistors' active area in tens of minutes. After rinsing the sample, we use a Zeiss Ultra 55 to acquire integrated circuit images similar to the

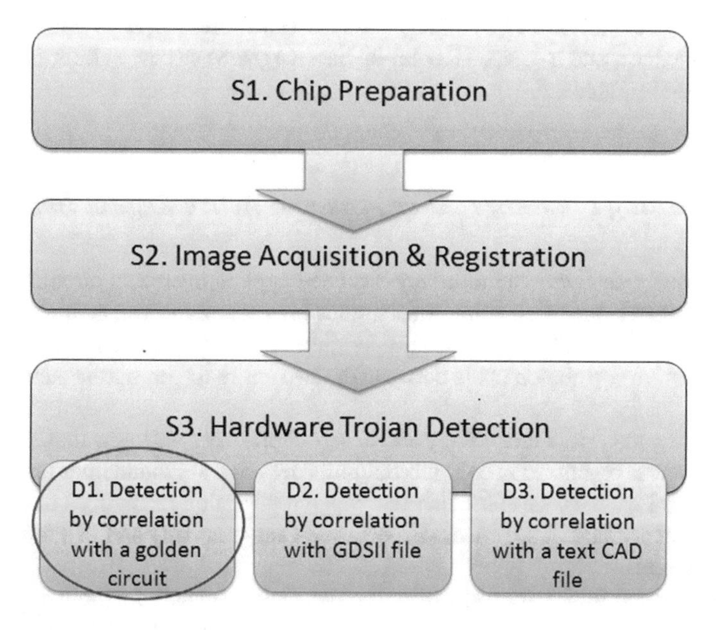

Fig. 7.13 Detection choice: golden model correlation

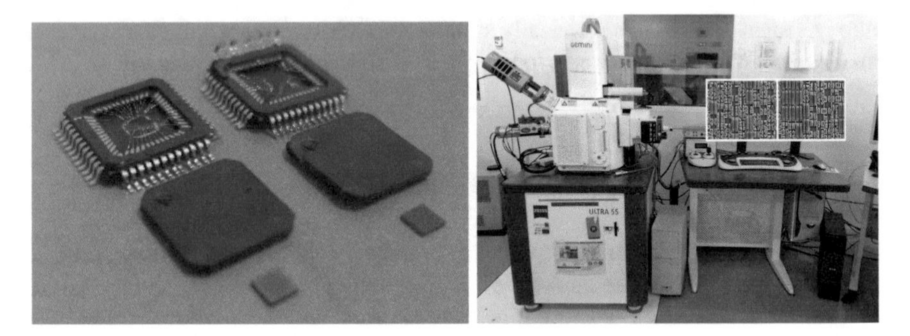

Fig. 7.14 *Left* Couple of ASICs being processed, *right* SEM platform

ones visible in the screens of the SEM in Fig. 7.14. We use open source algorithms to process acquired images.

7.7.2 The Hardware Trojan

The Hardware Trojan has been inserted at the mask level in GDSII format. It has been designed to be stealthy in terms of additional footprint and activity. The first asset of this HT resides in its number of additional gates. No extra Silicon has been used for the modification covering about 0.5 % of the integrated circuit. The HT trojan uses about 190 gate equivalent (GE). The inserted Hardware Trojan is combinational, 30-bits sequence, and triggers a Denial-of-Service (DoS) as payload. Its features are summarized in Table 7.5.

7.7.3 Getting the Entire IC Transistors' Active Region Image

The previously depicted methodology has been applied over both circuits. We first show the importance of cleaning the samples after using the wet etching bath (HF acid) in Fig. 7.15. It allows similar preparation results for both circuits and hence helps the future correlation made between the two entire IC transistors' active region images.

According to CMOS technology and designer practices we know that a chip synthesized logic is organized in columns delimited by vcc and ground lines linked to the substrate. In Fig. 7.16, we show that the image registration result does not give any inconvenient artifacts. The process is performed automatically and in a few minutes for a full circuit.

The so-called entire IC transistors' active region images for both circuits are shown in Fig. 7.17. Approximatively 20 min were required to obtain each image,

Table 7.5 Properties of the Hardware Trojan present in the device under test

Design phase	Abstraction level	Activitation	Effects	Location	Physical characteristic
Specification	System level	Always on	Change function	**Processor**	Distribution
Design	Development environment	Internally time based triggered	Degrade perf.	Memory	Size
Fabrication	RT level	**Internally physical cond. triggered**	Leak information	I/O	Parametric type
Test	**Gate level**	Externally user triggered	**Denial-of-service**	Power supply	**Functional type**
Assembly and Package	Transistor level	Externally component triggered		Clock	Layout-same structure
	Physical level				**Layout-change structure**

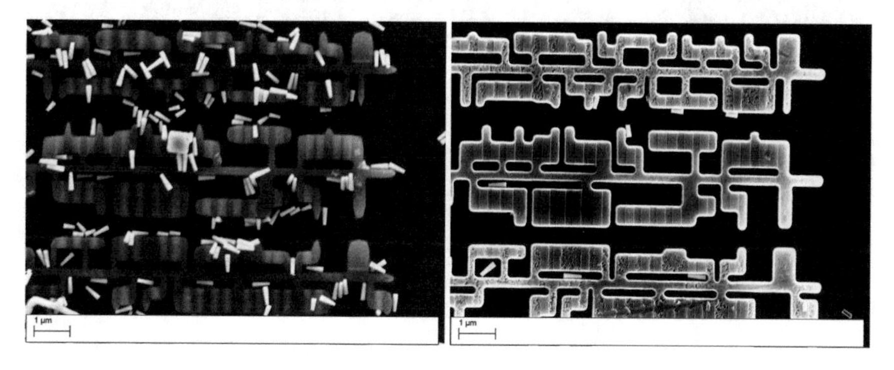

Fig. 7.15 *Left* A few transistors' active areas after wet etching, *Right* same area after ultrasonic bath application

respectively a set of '9 × 11' and '9 × 12' images. The magnification is 2k× and a secondary electron detector is used.

As seen in the simulated case, overlaying both images does not permit a straightforward way to find differences between them. Even being successive reconstructed acquisitions, images show intensity differences. Each sample is placed on a Carbon tape leading to a non-flat surface. Moreover, the sample holder can be tilted and the stage has to be translated and rotated to align the image of the second IC. Unlike optical microscopy, SEM imaging focus is almost non impacted over the full circuit area, the depth of field is much larger. However, it affects the matter/electron

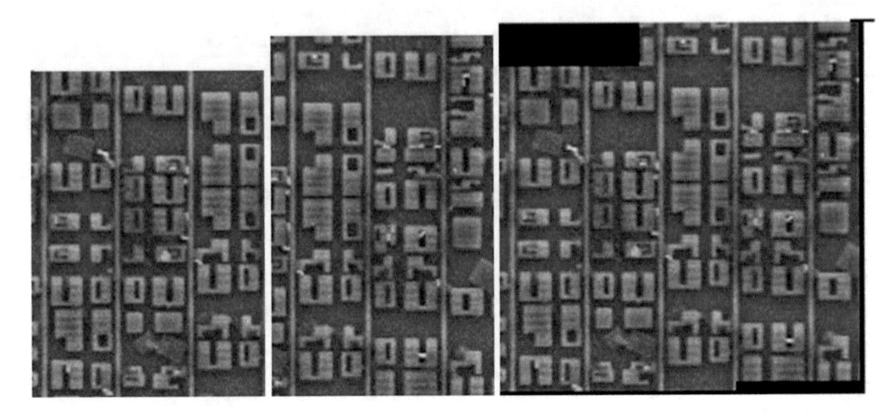

Fig. 7.16 *Left* First acquisition, *middle* second acquisition, *right* output aligned image

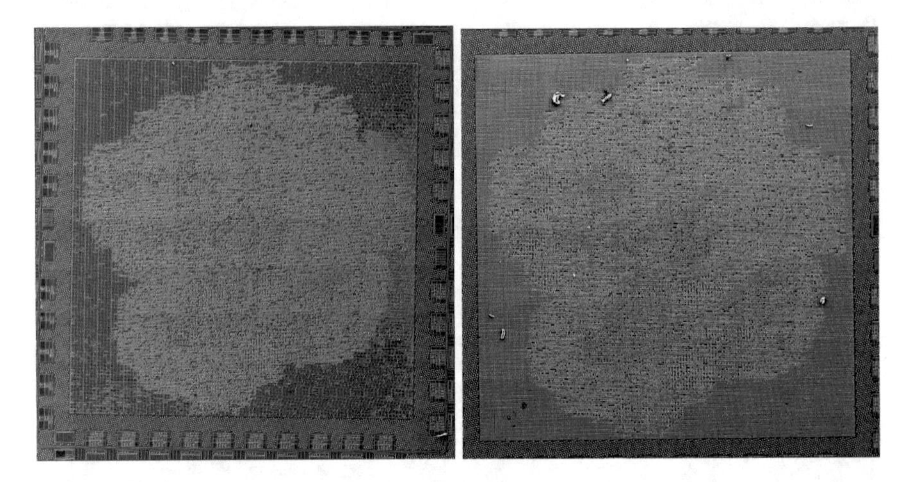

Fig. 7.17 SEM acquisitions covering the entire surface of genuine and infected ICs

interaction surface and thus the integrated circuit size actually seen on display and saved.

Adding the fact that the area to scan is manually defined twice, both final reconstructed IC images can not be directly compared.

7.7.4 Detecting the HT Knowing Its Location

We take advantage of the knowledge of the approximate location of the Hardware Trojan. Its location is shown in Fig. 7.18. Recall that the HT represents 0.5 % of the total number of standard cells used in the genuine circuit. The two images issued from each circuit preparation are manually aligned to highlight standard cell differences between them. For this manual registration, a simple image editor can be used. We manipulate the rotation, the translation, and the scaling of an image to compare with the other one. This manipulation enables us to perfectly align a small area (tens of standard cells only) and thus highlights changes brought to the design. However, the rest of the test circuit image is not correctly aligned with the genuine circuit image. Illustrated in Table 7.6 are drifts readings at the four extremities of the overlaid images. Over this table, misalignments are visible and correspond to a part of the south-west corner of the overlaid image (overlay done on HT area only which is more central). For instance, the same magnification over the south-west area also shows intensity variation for an identical element visualized in two different acquisitions, Table 7.6 (top right image).

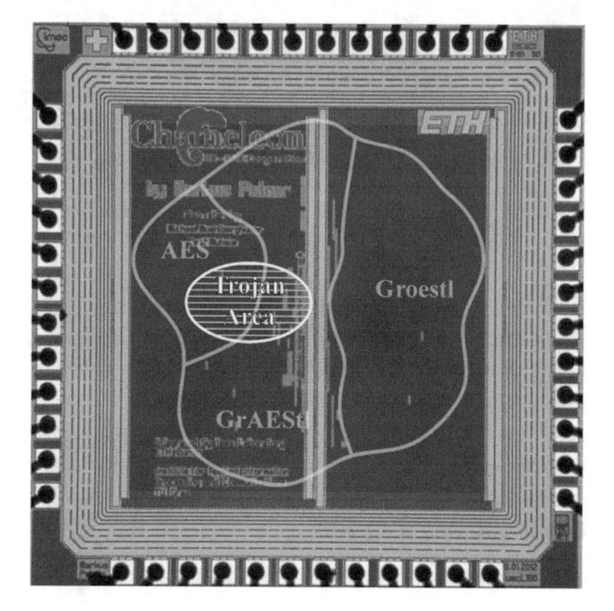

Fig. 7.18 Global *top view* of the integrated circuit with the area infected, figure extracted from [19]

Table 7.6 Observed drift during manual registration and overlay differences

Pixel info: (X, Y) Intensity

74	75	97	130	156	140	124	113	102
69	72	95	127	152	144	126	110	107
72	74	96	133	156	150	130	115	109
95	93	116	156	177	165	151	135	126
152	149	172	211	236	224	211	203	200
153	150	170	210	238	228	216	212	210
157	153	166	198	227	223	214	212	211

Localization	X drift in *pix.*	Y drift in *pix.*
North-West	-6	-2
South-West	-9	-10
North-East	15	19
South-East	12	16

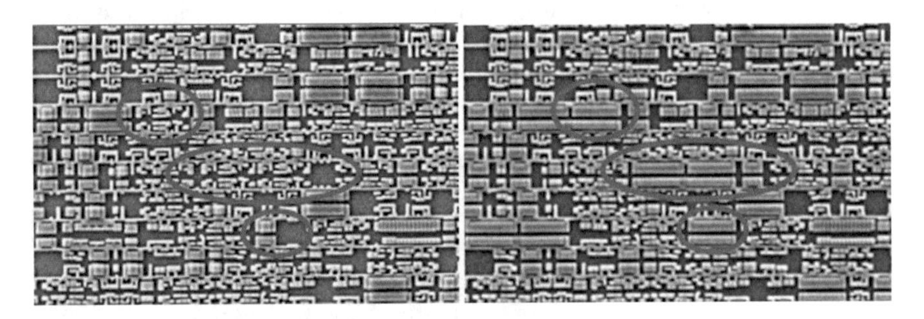

Fig. 7.19 *Left* Infected circuit HT area, *right* genuine circuit identical area

Besides detecting whether the test circuit is genuine, the technique also allows us to directly see the elements that have been modified, Fig. 7.19. Over those test circuits, filler cells have mostly been replaced by shapes whose characteristics (appearances, size) give the hint of flip-flop cells.

To sum up, the proposed methodology allows us to retrieve modified cells over the device despite the use of two physically different integrated circuits and two different image acquisitions. Knowing the location of the HT, we validate this approach to detect Hardware Trojans.

In a real life scenario, we admit that the HT presence is not communicated and its location even less known. Hence, the next subsection deals with detecting HTs without any a priori information.

7.7.5 *Detecting HT Invariant of Their Location*

This part of the methodology is based on a script that can be reused independently on the images pair. This is a semiautomatic process as we first manually select similar points over both images, then, based on a given algorithm, we recover transform coefficients canceling rotation, translation, and scaling differences between both images. We apply those coefficient weights to the second IC image. We then save the modified second IC image and load both images under a common image editor. Both images are finally overlaid, Fig. 7.20. Differences between the genuine and the infected integrated circuit are thus highlighted.

Table 7.7 gathers all the different methodology steps for this golden circuit detection application. The necessary operating/developing time is also given to highlight one of the main advantages of our proposed methodology, the small amount of time required.

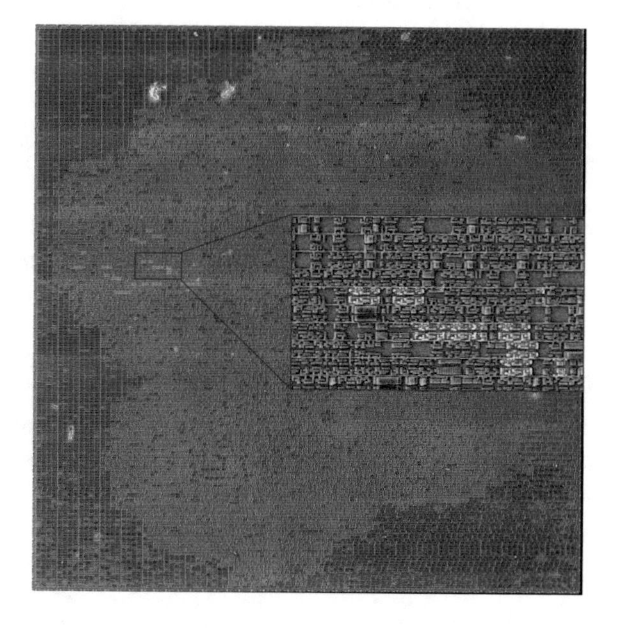

Fig. 7.20 Final image obtained, zoomed in to the HT area

Table 7.7 Hardware Trojan detection global process

Tasks	Role	Operating time/(development time)
Metal layer removal	Accessing the BULK layer	10 min/(0 min)
Cleaning bath	Removing wet etching residues	10 min/(0 min)
SEM parameters set	Imaging standard cell active regions	15 min/(0 min)
Full IC #1 SEM acquisition	Automated acquisition process for IC #1	20 min/(15 min)
Full IC #2 SEM acquisition	Automated acquisition process for IC #2	20 min/(NA)
Multiple Image Alignment	Registering each IC subparts	3 min/(20 min)
Reference & DUT images registration	Preparing the final overlay	2 min/(20 min)
Final images overlay & DUT images registration	Highlighting the HT presence	1 min/(2 min)

7.8 Conclusion and Perspectives

7.8.1 Conclusion

The demand for integrated circuits free of Hardware Trojan will continue to increase in the future. Regarding this threat, we propose a novel partial hardware reverse engineering methodology. The methodology includes sample preparation, image acquisition and image processing to detect any malicious hardware modification. First, sample preparation allows reaching transistors' active regions. Second, Scanning Electron Microscopy (SEM) permits to automatically acquire transistors' information over the full circuit. At last, image processing allows registering subset of IC images and detecting any difference compare to a genuine device. The proposed methodology has advantages in terms of efficiency, speed, cost, process and setup variation robustness, technology node robustness, yield impact and manufacturing flow compliancy. This technique has been successfully validated on real ICs. In less than an hour and with less than 200 Euros of equipment when renting, a small combinational Hardware Trojan (few dozens of gates) is successfully and almost automatically detected within a 40k gates IC.

7.8.2 Perspectives

Future investigations could involve:

- Non-destructive Hardware Trojan detection: an infrared camera could be used to see through the IC's Silicon substrate. It would be interesting to have enough resolution to compare the genuine IC with the noninfected one,

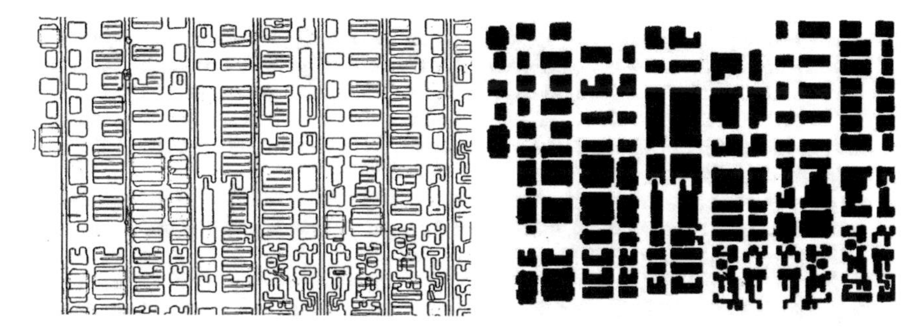

Fig. 7.21 Further image processing, *left* binary image, *right* filled shape image

- ECO cells or sub-active region HT detection: modifying or adding a step to the proposed methodology would be interesting to check Engineering Change Orders (ECO) cells connections and sub-dopant status,
- HT detection practical application without hard-golden: as stated in this chapter, a CAD-tool output file (such as GDSII or DEF file) could be used as golden reference,
- applying more image processing techniques over the entire IC transistors' active region image as shown in Fig. 7.21. Indeed, only the shape of a cell could be of interest. Shape statistics could also be sorted out to gain information over the circuit design.

Acknowledgements This work has been done within the HOMERE project funded by the French Government (BPI-OSEO) under grant FUI#14. This work has also been funded by the ANRT CIFRE funding #2012–2008. It leads to the obtention of my Ph.D. entitled 'Partial hardware reverse engineering for 'fine-grained' laser fault injection and efficient hardware trojan detection'. All my investigations could not have been realized without my industry supervisor Philippe Loubet-Moundi, my academic supervisor Jacques J. A. Fournier and my Ph.D. director Assia Tria, thank you so much. I would also like to thank my industrial team, my academical team and people in charge of cleanroom facilities. Many thanks to the ETH Zurich Integrated Systems Laboratory for making samples available to us, http://asic.ethz.ch/cg/2011/Chipit.html. Last but not least, acknowledgment to Jeunese Payne and David Llewellyn-Jones for proofreading and more generally to the University of Cambridge Computer Laboratory to let me the time to write this chapter.

References

1. M. Tehranipoor, F. Koushanfar, A survey of hardware trojan taxonomy and detection. IEEE Des. Test Comput, **27**(1), 10–25 (2010)
2. S. Blythe, B. Fraboni, S. Lall, H. Ahmed, U. de Riu, Layout reconstruction of complex silicon chips. IEEE J. Solid-State Circuits **28**(2), 138–145 (1993)
3. N.G. Bourbakis, A. Mogzadeh, S.J. Mertoguno, C. Koutsougeras, A knowledge-based expert system for automatic visual vlsi reverse-engineering: VLSI layout version. IEEE Trans. Syst. Man Cybern. Part A **32**(3), 428–436 (2002)

4. D. Lagunovsky, S. Ablameyko, M. Kutas, Recognition of integrated circuit images in reverse engineering, in *International Conference on Pattern Recognition*, vol. 2, 1998, pp. 1640–1642
5. G. Masalskis, R. Navickas, Reverse engineering of CMOS integrated circuits (2008)
6. R. Chakraborty, S. Narasimhan, S. Bhunia, Hardware Trojan: threats and emerging solutions, in *High Level Design Validation and Test Workshop, 2009. HLDVT 2009. IEEE International*, 2009, pp. 166–171
7. C. Bao, D. Forte, A. Srivastava, On application of one-class SVM to reverse engineering-based hardware Trojan detection, in *2014 15th International Symposium on Quality Electronic Design (ISQED)*, 2014, pp. 47–54
8. S. Bhasin, J.-L. Danger, S. Guilley, X. Ngo, L. Sauvage, Hardware Trojan horses in cryptographic ip cores, in *2013 Workshop on Fault Diagnosis and Tolerance in Cryptography (FDTC)*, 2013, pp. 15–29
9. R. Torrance, D. James, The state-of-the-art in ic reverse engineering, in *CHES 2009*, pp. 363–381
10. R. Kumar, P. Jovanovic, W. Burleson, I. Polian, Parametric Trojans for fault-injection attacks on cryptographic hardware. Cryptology ePrint Archive, Report 2014/783 (2014), http://eprint.iacr.org/
11. https://www.commoncriteriaportal.org/files/supdocs/ccdb-2013-05-002.pdf, (2013)
12. F. Courbon, J.J.A. Fournier, P. Loubet-Moundi, A. Tria, Combining image processing and laser fault injections for characterizing a hardware AES. IEEE Trans. CAD Integr. Circuits Syst. **34**(6), 928–936 (2015), doi:10.1109/TCAD.2015.2391773
13. F. Courbon, Partial hardware reverse engineering applied to fine grained laser fault injection and efficient hardware Trojans detection. Theses, Ecole Nationale Supérieure des Mines de Saint-Etienne (2015), https://tel.archives-ouvertes.fr/tel-01258054
14. F. Beck, Integrated Circuit Failure Analysis: A Guide to Preparation Techniques. ser. Quality and Reliability Engineering Series (Wiley, 1998), http://books.google.fr/books?id=7VNfvKjlzYAC
15. H. Kaeslin, *Digital Integrated Circuit Design: From VLSI Architectures to CMOS Fabrication*, 1st edn. (New York, USA, 2008)
16. F. Courbon, P. Loubet-Moundi, J.J.A. Fournier, A. Tria, Increasing the efficiency of laser fault injections using fast gate level reverse engineering, in *2014 IEEE International Symposium on Hardware-Oriented Security and Trust, HOST 2014, Arlington, VA, USA, May 6–7, 2014*, 2014, pp. 60–63, doi:10.1109/HST.2014.6855569
17. J.J.A. Fournier, J. Rigaud, S. Bouquet, B. Robisson, A. Tria, J. Dutertre, M. Agoyan, Design and characterization of an AES chip embedding counter measures. IJIEI **1**(3/4), 328–347 (2011), doi:10.1504/IJIEI.2011.044101
18. J. Lewis, Fast normalized cross-correlation. Vis. Interf. **10**(1), 120–123 (1995)
19. M. Muehlberghuber, F.K. Gürkaynak, T. Korak, P. Dunst, M. Hutter, Red team vs. blue team hardware Trojan analysis: detection of a hardware Trojan on an actual ASIC, in *HASP 2013, The Second Workshop on Hardware and Architectural Support for Security and Privacy, Tel-Aviv, Israel, June 23–24, 2013*, 2013, p. 1, doi:10.1145/2487726.2487727

Chapter 8
Linear Complementary Codes: Novel Hardware Trojan Prevention and Detection Approach

Xuan Thuy Ngo, Sylvain Guilley and Jean-Luc Danger

8.1 Introduction

The semiconductor industry has spread across borders in the time of globalization. Different design phases of an integrated circuit (IC) may be performed at geographically dispersed locations. This coupled with the outsourcing design and fabrication to increase profitability has become a common trend in the semiconductors industry. However, this business model comes with an ample scope of introducing malicious behavior to a part of the IC. The adversary has enough scope to tamper the supply chain by maliciously implanting extra logic as hardware Trojan horse (HT) circuitry into an IC [30]. This raises serious concerns about security and trustworthiness of imported products employed in critical applications like military, health, transportation, etc. HT can be introduced in an IC at several points right from the register transfer level (RTL) source code to lithographic masks fabrication. An attacker can change a design netlist or subvert the fabrication process by manipulating design masks, without affecting the main functionality of the design [2].

Any HT is composed of two main components [29]:

- **Trigger**: is the part of HT used to activate the malicious activity,
- **Payload**: is the part of HT used to realize/execute the malicious activity.

Figure 8.1 shows an example of one simplistic HT. In this archetype HT, the trigger is a simple AND gate: it tests the equality of the inputs A and B to 1; the payload is an XOR gate: it inverts the intermediate net C if the trigger is active.

An adversary can introduce a HT which might be designed to disable or destroy a system at some future time, or to leak confidential information such as secret keys

X.T. Ngo (✉) · S. Guilley · J.-L. Danger
Telecom Paristech, 46 rue Barrault, 75013 Paris, France
e-mail: xuanthuy.ngo146@gmail.com

S. Guilley · J.-L. Danger
Secure-IC SAS, 5510 Cesson-Sevigne, France

L. Bossuet and L. Torres (eds.), *Foundations of Hardware IP Protection*,
DOI 10.1007/978-3-319-50380-6_8

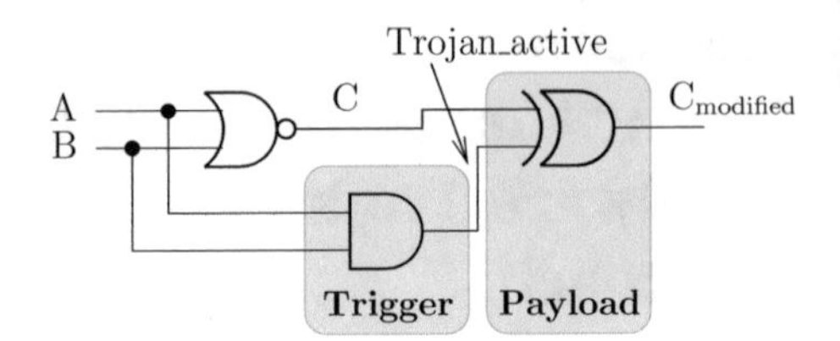

Fig. 8.1 Minimalist hardware Trojan horse example

covertly to an adversary [28] by putting them to output channels. In [27], the author demonstrates an attack on a (purported military-grade) chip using a malicious backdoor. The backdoor allows the attacker to disable all the security of the chip, reprogram cryptographic part, access secret keys, modify low-level silicon features, access unencrypted configuration bitstream, or permanently damage the device. Thus protection against HT is an open problem and an active research topic.

HT detection is an extremely challenging problem; traditional structural and functional tests do not seem to be effective in targeting and detecting HTs. Since HT can be introduced during different design phases, the nature of HT differs from one design phase to the others. Therefore, it is difficult to find a unique detection technique for all HT. For instance, automatic test pattern generation (ATPG) methods which are used in manufacturing test for detecting defects generally operate on the netlist of the HT-free circuit. Existing ATPG algorithms cannot target HT activation/detection directly [31] because HTs are designed such that they are silent most of their lifetime and have very small size relative to their host design, with featuring limited contribution into design characteristics. Such HTs are most likely connected to nets with low controllability and/or observability [6, 31].

The state-of-the-art principle in detection strategies of HT can be widely classified into two categories, *viz.* invasive and noninvasive [15].

Noninvasive HT detection is done by comparing the performance characteristics of an IC with a known good copy also known as the "golden circuit." Detecting HT in a noninvasive manner can be done either at runtime or in the testing phase. The runtime detection mechanism is combined with the countermeasures, as once a HT is detected at runtime, attempts are made to try and continue operating by bypassing the HT. For runtime, Bloom et al. detail a HT detection approach that uses both hardware and software to detect two types of HT which are DoS (Denial of Service) and combined hardware and software HT [7]. Abramovici and Bradley added reconfigurable DEsign-For-ENabling-SEcurity (DEFENSE) logic to the functional design to implement real-time security monitors [1]. McIntyre et al. detect the presence of HTs by executing functionally equivalent processes on multiple hardware processing elements [21]. The testing phase detection methods attempt to enhance traditional IC testing, or use side-channel analysis. For logic testing, Jha and Jha present a randomisation-based technique which probabilistically compares the functionality of the design of the circuit with the implemented circuit [18]. Chakraborty et al. suggest to test rare occurrences on an IC rather than testing for correctness [11]. The tester determines rare states that can occur within a circuit module. For side-channel analysis, Agrawal et al. present a type of detection mechanism [3]. Some known good copies of the IC are obtained and fingerprinted using one or more side-channel parameters. Other chips can then be tested against these fingerprints like path delay in [19]. Power supply transient signal analysis is used as the side channel by Rad et al. [24]. They aim at determining the smallest HT that they can find using this technique, which can be as low as three additional gates. Banga and Hsiao [6] propose the "sustained vector technique" that is able to magnify the side-channel differences (based on power draw) between circuits infected with HTs and those that are not.

Invasive methods try to (prophylactically) modify the design of IC to prevent the HT or to assist another detection technique. One prevention method against HT has been presented by Chakraborty et al. at ICCAD 2009 [9]. It is inspired by obfuscation methods [4, 10] initially intended to protect against IC counterfeiting. In this paper, authors obfuscate the original design by increasing the total number of reachable states of the original circuit. These states are partitioned into two parts: an original state space and an isolation state space. The original state space will be reached using a specific input pattern (as a secret key). Moreover, with any wrong input pattern, the IC will fall in the isolation state space. This space is constructed such that, once entered, it cannot be exited and outputs will never be correct.

Another technique, nicknamed ODETTE [5], aims at changing the polarity of the flip-flops (also known as DFFs). This option can be achieved at low cost, since DFFs of standard libraries (provided by the founders) feature two complementary outputs (called $\overline{Q}$ and Q). This coding is akin to Vernam cipher, where each bit of the state is masked with one bit of secret. Authors claim that it is able to obfuscate partially the circuit. In [12], authors present a logic gates encryption technique using an external key to prevent HT insertion.

The drawback of these prevention methods in the state of the art is that they obfuscate only the state machine of the IC. This means that only the control part is protected, while the combinational part is unprotected. Moreover in papers presented in [9, 12], when the IC is well configured to reach the original state using static configuration keys, it operates normally and cannot resist others physical attacks. The prevention method, ODETTE [5], is more intended to raise the HT activity for a better detectability than a proactive prevention. Furthermore, each bit of the state is masked with one bit of secret.

In this chapter, we intend to find a more flexible solution, where the number of "mask" bits can be chosen, thus allowing the designer to adjust the security level. We propose the concept of **"encoded circuits"**, a provable randomization method using the linear complementary pair (LCP) codes C and D to prevent HT insertions. Encoded circuits are realized by encoding all internal registers (sequential part) of the target design with a *binary code* C and followed by addition (XOR) of random masks in its *supplementary code* D. Once the sequential part is encoded, the combinational part can be easily obfuscated by exploiting the "flatten" option of the netlist synthesis tool. It merges the logic part of encoder/decoder circuit with the combinational part of the target circuit. Thus, the state is totally encoded and the structure of original combinational part is totally lost when synthesized together with the encoder and decoder circuit. After encoding, the complexity of the design increases which obfuscates the real functionality of the IC. Using our coding method, we manage, to some extent, to protect both control and data parts. Moreover, we can not only protect against HT insertion attack but also against other physical attacks because of the use of random masks.

Our technique is closely based on the *private circuits* of Ishai, Sahai, and Wagner presented at CRYPTO 2003 [17]. Private circuits were proposed to protect against probing attacks which ensures no information leakage with $\leq d$ probes, d being a security parameter. In this chapter, we go even further by proposing two security parameters $d_{Trigger}$ and $d_{Payload}$. The parameter $d_{Trigger}$, which is the dual distance of code D, ensures that HT connected to less than $d_{Trigger}$ registers or nodes will not retrieve useful information. And the parameter $d_{Payload}$, which is the minimal distance of code C, ensures that any HT modifying less than $d_{Payload}$ will be detected. So **"encoded circuit"** is at the same time prevention and runtime detection method. $d_{Trigger}$ and $d_{Payload}$ parameters can be chosen independently in order to increase the prevention capacity or detection capacity. And the overhead will depend to these two parameters. In this chapter, we provide the rationale behind the encoded circuits. We first describe the theoretical background of encoded circuits based on theory of codes. Thereafter, we detail the techniques to choose and generate LCP codes for encoding a circuit. Then, we present the design flow for encoding method integration. Practical application of encoded circuits is demonstrated on a simple microprocessor, SIMON cryptographic coprocessor as well as an AVR processor. We show that the technique can be applied to any circuit to prevent HT insertion. After, we present how LCP codes could be optimized to reduce its hardware implementation overhead. In the end, we discuss about the efficient of encoded circuit method against other physical attacks.

The rest of this chapter is structured as follows. Section 8.2 gives the concept of encoded circuit method. Section 8.3 presents the definition of security parameters as well as the properties of LCP codes. Section 8.4 shows how LCP can be integrated automatically on design flow using scripts. Section 8.5 demonstrates case studies of LCP method on different circuits. Then the optimization algorithm of LCP is presented in Sect. 8.6. The performance of encoded circuits against other physical attacks as SCA and FIA are tested in Sect. 8.7. The comparison of LCP method with those in the state of the art is given in Sect. 8.8. Finally we conclude in Sect. 8.9, and give some perspectives.

8.2 Encoded Circuit Concept

In this section, we detail the rationale of encoded circuits. We first describe the basic principle of encoded circuits. Next, we define security objective which determines the choice of codes for encoded circuits.

As presented in Sect. 8.1, a HT can be globally seen as a composition of two part:

- **Trigger**: which reads the target circuit state (to trigger its malicious function).
- **Payload**: which writes on the target circuit state (to realize its malicious function).

One can observe from HT structure that to compute the good trigger value, an attacker must have knowledge of the circuit. In a complex circuit, a trigger must be connected to a number of nets (or I/O pads) in order to be controllable and efficient. If the HT depends on very few signals, then the HT activation rate increases, which makes it detectable. Therefore, HT acts as a "probing (or side-channel) station" which is built into the circuit.

The principle behind encoded circuit is very basic. Every IC is composed of two distinct kinds of logic cells: sequential (D flip-flop or register) and combinational. In practice, it is known that registers are easily recognizable, because they are much larger than combinational gates. For instance, in the 130 nm technology of ST-Microelectronics, the size of a D flip-flop (in short: DFF) varies from 30 to 50 μm^2, while INVERTER (IVLLX05), NOR (NR2LL), and NAND (ND2LL) sizes are, respectively, 4.03, 6.05, and 6.05 μm^2. A section of a layout is shown in Fig. 8.2 containing both combinational and sequential logic. The yellow blocks in Fig. 8.2 are DFF or sequential logic and the red blocks are combinational gates. We can easily notice that DFF is larger than combinational logic gates. DFF gates can be recognized directly in an IC. It can be seen Fig. 8.3 that DFF stands out clearly from combinational logic. Another motivation for probing the DFF output is that the

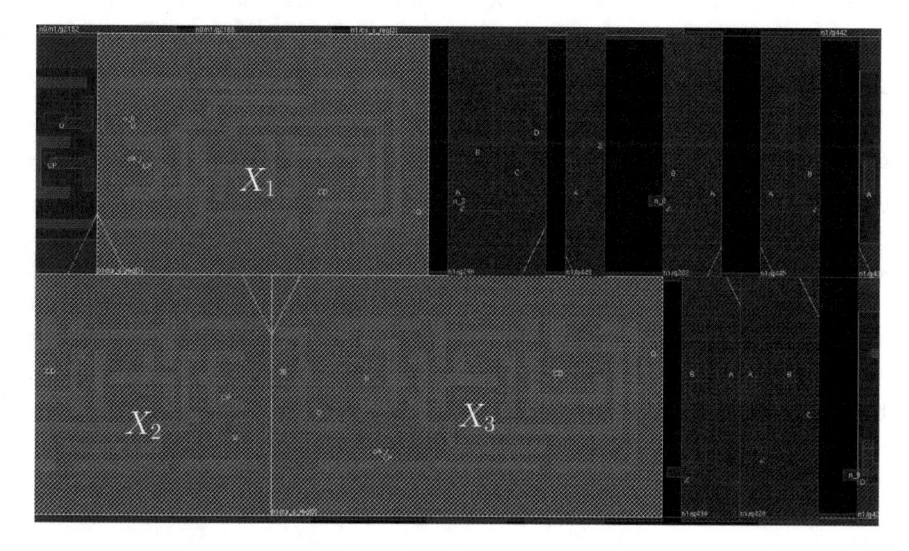

Fig. 8.2 Floor plan of an IC with D flip-flops (in *yellow*)

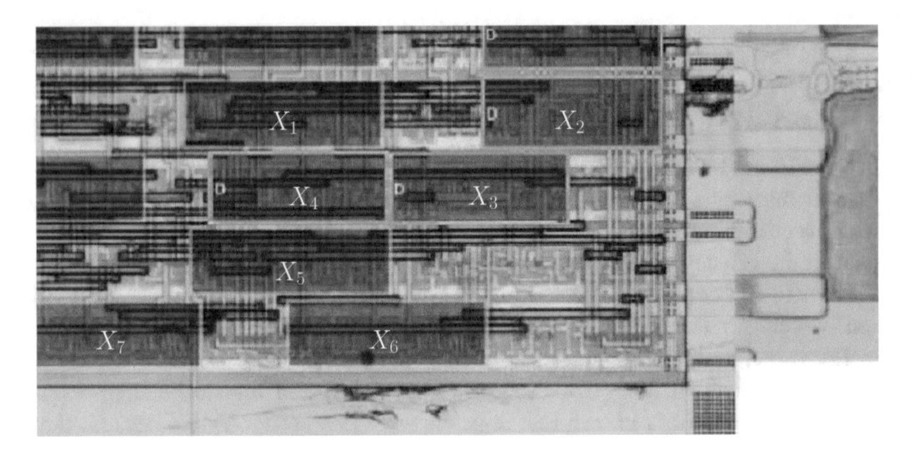

Fig. 8.3 Layout of an ASIC with D flip-flops [23] (in *gray*)

signals at DFF output are synchronized. Therefore, it is easier for an attacker to insert HT using the inputs or outputs of these registers as activation conditions. Thanks to this observation, we propose to apply the concept of "private circuits" [16, 17], which is used to encode mask all gates of the target circuit, only on internal registers. This protection is initially designed to resist the theft of probed signals. Our encoded circuit method goes beyond, insofar as it shall resist against more connection than a mere probing attack. Another specificity of our protection is that it impedes both the trigger and payload parts of a supposedly inserted HT. Using the encoding system, we can transform the original data of all sequential logic cells to the encoded and masked data, hence protecting them.

Let us call x the state, that is to say, the set of all sequential resources. We denote by k the number of state bits in the original circuit. It is encoded as follows:

- a code C of length n, which is applied on x. For the sake of simplicity, we assume that C is a linear Boolean code.
- some random numbers y of $(n - k)$ bits, which serve as a pool of entropy to mask the encoded state. The masks are also encoded, by a code D, of size $(n - k)$ and dimension n. As a result, the encoded and masked state z is the exclusive-or of one code word of C and D each.

If C and D are supplementary, the encoded x and y can be retrieved from z. We denote by G and H the generator matrices of C and D. It is thus required that the $n \times n$ square matrix $\begin{pmatrix} G \\ H \end{pmatrix}$ be of full rank n.

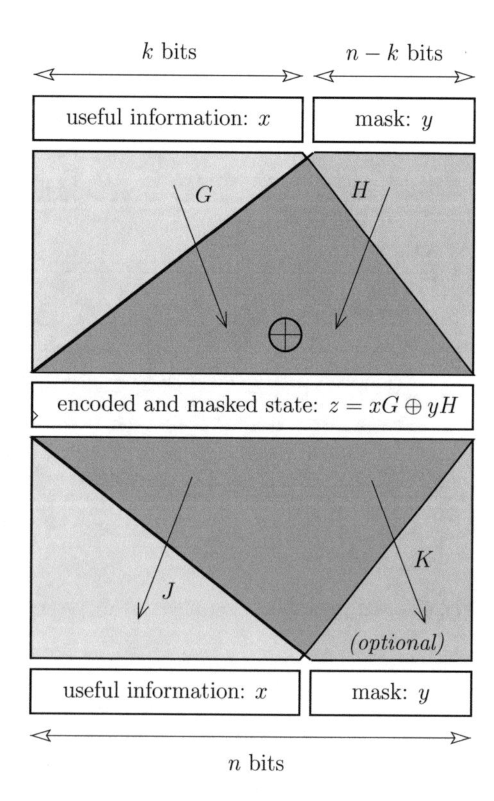

Fig. 8.4 Principle of "circuit encoding"

The decoding logic allows to recover x from z. This operation is also a linear function that maps elements of $\mathbb{F}_2^n$ to $\mathbb{F}_2^k$ (i.e., not injective). And we denote by J and K decoding generating matrices of C and D.

Next, we can also check the random numbers which belong to the code D. If the state z is corrupted by some means, that would also impact x and y. Therefore, it is relevant to recover y from z, which can be done by a linear function of generator matrix K.

The encoding/decoding functions are summarized in Fig. 8.4. Unless otherwise mentioned, we will assume in the sequel that C and D are supplementary. We denote supplementary pair codes, such as C and D, "LCP" (for "Linear Complementary Pair"). This term has been coined by Massey in [20]. A full schematic of the encoding principle is illustrated in Fig. 8.5; added blocks are represented in blue color.

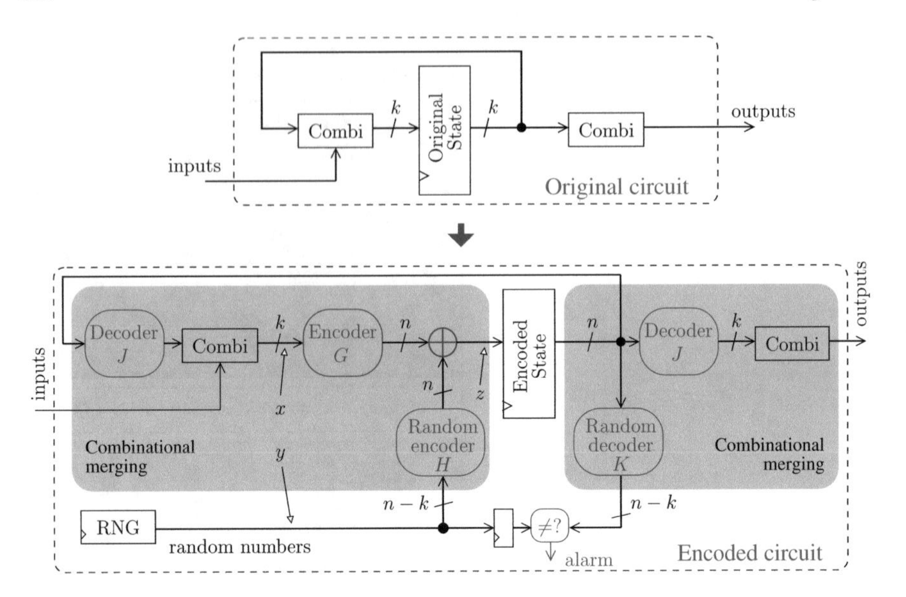

Fig. 8.5 Architecture of "Encoded Circuit," exemplified on a canonical Moore machine

8.3 Encoded Circuit Method Properties

8.3.1 Security Objective

We intend to apply the notion of private circuits, discussed by Ishai, Sahai and Wagner at CRYPTO 2003 (see [17]). They introduce a security metric d. Let d be an integer. They design a method to realize circuits such that probing any tuple of strictly less than d equipotentials does not allow to derive any information about the real data manipulated by the circuit. Their construction has a quadratic complexity in d. Specifically, if the circuit has initially k gates, the size of the encoded & masked circuit they propose is $\mathcal{O}(k \cdot (d+1)^2)$.

In our construction, we focus on the encoding of the k registers (there are much less registers than combinational gates in most circuits). This simplifies the problem, as the functionality of a register is simply the "identity function," which is linear. This is why linear codes are suitable in our case, which will further result in lower complexity of the protected circuit.

Specifically, in our method, we apply a linear complementary pair codes C and D. A pair of codes C and D is LCP if they are complementary. We define the minimal distance of C and the dual distance of D as 2 distinct security parameters d_{Payload} and d_{Trigger}.

A characterization of the two parameters of LCP codes is the following:

- $d_{\textbf{Trigger}}$: insures that HT, which probes $d_{\text{Trigger}} - 1$ (or less) bits of the encoded & masked state z, does not disclose any information on x.

- $d_{\mathbf{Payload}}$: insures that HT, which modifies $d_{\text{Payload}} - 1$ (or less) bits of the encoded & masked state z, cannot produce a valid codeword.

This is feasible, as stated in the following properties.

Property 1 *The encoding of x as $z = xG \oplus yH$, where y is a uniformly distributed mask in $\mathbb{F}_2^{n-k}$, does not reveal any information on x provided up to $d_{Trigger} - 1$ bits of z are known, if and only if D is of dual distance $d_{Trigger}$.*

Proof The mask y is applied additively on the encoded state xG as yH. The property that is required is that any tuple of size strictly less than d be balanced. As y is assumed uniformly distributed in $\mathbb{F}_2^{n-k}$, the distribution of any such tuple is unchanged (hence uniform) if and only if the code D is of dual distance d_{Trigger} (or more).

Property 2 *Let us consider the encoding of x as $z = xG \oplus yH$, where y is a uniformly distributed mask in $\mathbb{F}_2^{n-k}$. Any fault on z of Hamming weight strictly smaller than $d_{Payload}$ can be detected, if and only if C is of minimal distance $d_{Payload}$.*

Proof The state z is modified into $z \oplus \varepsilon$, for some random $\varepsilon \in \mathbb{F}_2^n$. By supplementary of C and D, there exists a unique ordered pair $(e,f) \in \mathbb{F}_2^k \times \mathbb{F}_2^{n-k}$ such that $\varepsilon = eG \oplus fH$. A detection strategy consists in checking whether or not the mask has been altered, i.e., $zK \stackrel{?}{=} y$. This verification does not jeopardize the security model of Property 1 since x is not uncovered, only y. By linearity of the fault injection, the equality $(z \oplus \varepsilon)K = y$ happens if and only if $\varepsilon K = 0 \iff f = 0$, i.e., $\varepsilon \in C$. As $\varepsilon = 0$ is pointless (since without observable effect), harmful (since undetected) faults only happen if and only if $\varepsilon \in C\backslash\{0\}$. In particular, a necessary condition for the fault to be undetected is that the Hamming weight of ε be greater than or equal to the minimal distance d_{Payload} of code C.

Now, given that the minimal distance d_{Payload} of C and the dual distance d_{Trigger} are a security parameter, they are set as high as possible. Therefore, have LCP codes C and D of greatest possible minimal distance and dual distance simultaneously improves the resistance against HT insertion and FIA?

Now, how to determine the parameters d_{Trigger} and d_{Payload}? The rationale is the following: large HT can be detected by various means (optical inspection of the chip, SCA, etc.). So, the minimal size of a HT that would be difficult to identify is captured by distances d_{Trigger} and d_{Payload}. A stealthy HT below those distances would have uncontrollable trigger and would certainly be captured red-handed when executing its payload.

Using the LCP codes, the sequential part of a circuit can be well encoded. Nevertheless, it is still possible to insert a HT. An attacker that can isolate all blocks of an encoded IC (i.e., combinational part, encoder data G, encoder noise H, and decoder data J) can bypass the prevention by inserting a HT which probes directly at the inputs of encoder block (or at the output of decoder block, etc.). This is all the more possible as the IC is synthesized (i.e., generated) hierarchically. Using the *flattening* option for the netlist synthesis, we will merge these blocks together (combinational part, encoder for data, encoder for noise and decoder for data) for protecting the combinatorial part of encoded circuit. Therefore, it becomes a challenge for an attacker to reverse the real functionality of the IC for HT insertions.

8.3.2 LCP Code Properties

For the construction of LCP codes, we need to create two space vectors C and D (seen as linear codes) that are supplementary, i.e., $C \oplus D = \mathbb{F}_2^n$, with those additional constraints:

1. D must be of dual distance d_{Trigger};
2. C must be of minimal distance d_{Payload}.

In our application, C is used to encode original state of k bits; therefore, the dimension of C is k and the dimension of D is $n-k$. So, for a given k, we search for the smallest $n \geq k$ such that

1. there exists a code D of parameters $[n, n-k]$ and of dual distance d_{Trigger}, i.e., there exists a code $C' = D^{\perp}$ of parameters $[n, k, d_{\text{Trigger}}]$,
2. there exists a code C of parameters $[n, k, d_{\text{Payload}}]$.

We write the generating matrix G of C in a systematic form $G = \left(I_k | M \right)$, where M is a $k \times n-k$ matrix. Similarly, we write the generating matrix H of D as $H = \left(N | I_{n-k} \right)$, where N is a $(n-k) \times k$ matrix.

Proposition 3 *The three following statements are equivalent:*

1. *The matrix* $\left(\frac{G}{H} \right) = \left(\begin{array}{c|c} I_k & M \\ \hline N & I_{n-k} \end{array} \right)$ *is invertible;*
2. *The matrix* $I_k \oplus MN$ *is invertible.*
3. *The matrix* $I_{n-k} \oplus NM$ *is invertible.*

Corollary 4 *When it is invertible (see Proposition 3), the inverse of matrix* $\left(\begin{array}{c|c} I_k & M \\ \hline N & I_{n-k} \end{array} \right)$ *is given by*

$$\left(\begin{array}{c|c} I_k & M \\ \hline N & I_{n-k} \end{array} \right)^{-1} = \left(\begin{array}{c|c} (I_k \oplus MN)^{-1} & M(I_{n-k} \oplus NM)^{-1} \\ \hline N(I_k \oplus MN)^{-1} & (I_{n-k} \oplus NM)^{-1} \end{array} \right).$$

Remark 5 There is one particular case where the codes C and D are orthogonal. In this time, the minimal distance of C is also the dual distance of D. It means that $d_{Trigger} = d_{Payload}$. Indeed, in this case, H is the parity check matrix of code C, i.e., $GH^{\mathsf{T}} = 0$, or equivalently, $HG^{\mathsf{T}} = 0$, and there exists an orthogonal projection. It can be checked that $J = G^{\mathsf{T}}(GG^{\mathsf{T}})^{-1}$. If $z = xG \oplus yH$, then $zG^{\mathsf{T}} = xGG^{\mathsf{T}} \oplus yHG^{\mathsf{T}} =$

$x(GG^\mathsf{T})$, which simplifies to $zG^\mathsf{T}(GG^\mathsf{T})^{-1} = x$. Indeed, let $z = xG \oplus yH$. Then $zG^\mathsf{T} = xGG^\mathsf{T} \oplus yHG^\mathsf{T} = x(GG^\mathsf{T})$, hence $zG^\mathsf{T}(GG^\mathsf{T})^{-1} = x$. And $K = H^\mathsf{T}(HH^\mathsf{T})^{-1}$. In this time, LCP codes become linear complementary dual (LCD) codes. And in the state of the art, there are several methods for generating LCD codes [20] (for example Quadratic-Residu Codes, ReedSolomon codes etc.).

8.4 Automated Design Flow for Encoded Circuit

We briefly described the theory of encoded circuits in previous section. The method to encode a standard digital circuit is straightforward, which makes it easy to automate. The fully automated design flow for encoding a given hardware to protect against HT insertion is shown in Fig. 8.6. The flow can be divided into six distinct steps which are as follows:

Logic Synthesis This step is native to any design flow. The user synthesizes a HDL description of the design with a synthesis script (in TCL), which constraints the tool to flatten the netlist (e.g., "`ungroup -flatten -all`" in Encounter RC from Cadence). This step ensures that we enter into the paradigm of the Moore machine such that all sequential elements are gathered into a global state. Next, we check that the design is coded in a way such that there is no logic from the clock and reset inputs till the flip-flops. Finally, the synthesizer is constrained not to use flip-flops that "compute," e.g., flip-flops with an enable or two inputs (it is usual to find these gates in standard cell libraries, because they are dedicated to the *test* of the circuit). So, we use only non-test flip-flops which can be enforced by the "`set_attribute avoid true libcell libcell_location`" TCL constraint in Cadence Encounter Compiler. The synthesis exports the netlist as `design.v`.

Split Design This is the first step of modified design flow: We identify and separate the sequential part of the design from the combinational part. For the sequential part, it is also important to keep the initial value at reset for each flip-flop. The final wildcard comprises various loads. The number of DFF is k, and their initial state is denoted as $x_0 \in \mathbb{F}_2^k$. For the combinational part of the circuit, it is sufficient to remove all the flip-flops, followed by addition of a `from_seq` input and a `to_seq` output

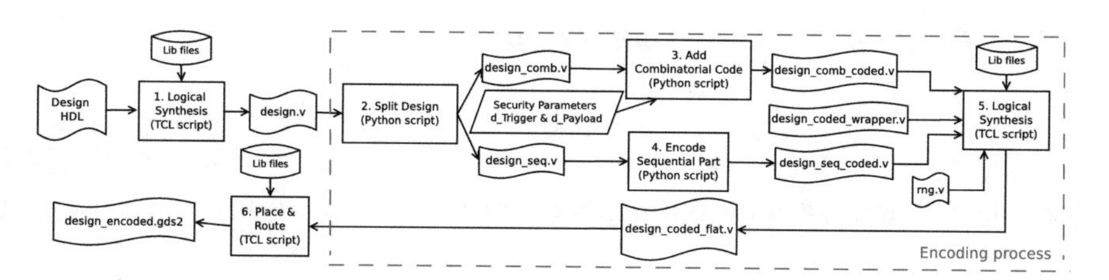

Fig. 8.6 Design flow for encoding method integration

bus (of bitwidth k). The automation is achieved with Python. This step generates two files: `design_comb.v` and `design_seq.v`.

Add Combinational Code In this step, the user inputs the security parameters d_{Trigger} and d_{Payload}. Using the value k derived from the previous step, the script generates HDL code for matrices G, H, J, and K for a suitable n. Next the file `design_comb.v` is connected with the HDL of matrices G, H, J, and K as shown in Fig. 8.5. The connection between matrices and the combinational circuit is done automatically using a Python script. This step generates `design_comb_coded.v` at the output and the hierarchical structure of the file is kept intact at this stage. Also, at this stage, the random number generator (RNG) which produces $(n - k)$ bits of random numbers at every clock period is considered as a black-box.

Encode Sequential Part The input of this step is `design_seq.v`. This step comprises regeneration of with data input/output as a bus of bitwidth n, and programmed with encoded initial state x_0G at reset. In other words, k flip-flops in the uncoded state are replaced by n flip-flops in the encoded state, keeping the equivalent state at reset.

Synthesize Encoded Design This step takes netlists `design_comb_coded.v`, `design_seq_coded.v` along with a RNG description `rng.v` and a wrapper circuit `design_coded_wrapper.v` as inputs. The function of the wrapper circuit is to connect the combinational and sequential part of the encoded circuit, while keeping the same interface as original design. All the files are fed to a logic synthesizer to generate a flattened netlist of the encoded design, i.e., `design_coded_flat.v`.

Place and Route The rest of the design flow is same as the standard design flow. In this step, the designer gives the synthesized netlist `design_coded_flat.v`. The design is then placed and routed to generate the final layout (GDSII).

8.5 Case Studies

In this part, we apply the encoded circuit method on three test circuits: Nanoprocessor, SIMON processor, and AVR processor.

8.5.1 Case Studies I: Nanoprocessor

For the first experiment, we choose nanoprocessor [32], which is a 8-bit processor without pipeline and requires 3 clock cycle to execute every instruction. It has 16 basic instructions, and operates using an external 256 bytes memory.

Table 8.1 Synthesis results of nanoprocessor-encoded circuit method, and security parameters

IC (Code)	Gates	Area (μm^2)	n	k	$d_{Trigger}$	$d_{Payload}$
Original ([37,37,1,1])	199	1181	37	37	1	1
Encoded ([73,37,13,13])	1001	6926	73	37	13	13
Encoded ([86,42,17,17])	1410	9717	86	37	17	17
Encoded ([89,45,17,17])	1754	11296	89	37	17	17
Encoded ([73,37,17,12])	1159	7377	81	37	17	12
Encoded ([73,37,17,8])	1151	7324	81	37	17	8
Encoded ([49,37,5,3])	433	3137	49	37	5	3

The unprotected nanoprocessor gives the following after synthesis:

- 37 sequential cells (flip-flops),
- 208 combinational cells.

Thus we have $k \geq 37$ for the nanoprocessor netlist.

First, we constructed the LCP codes with $d_{\text{Trigger}} = d_{Payload}$ for nanoprocessor. We apply a LCP codes **[73,37,13,13]**, i.e., $k = 37$, $n = 73$ [8]. Since $k = 37$ equals number of flip-flops in nanoprocessor, this is the smallest code which can be applied. The security parameter $d_{\text{Trigger}} = d_{\text{Payload}}$ of this code is 13, i.e., an attacker should connect or modify to at least 13 DFF to implement an effective HT. To achieve a larger security parameter, the dimensions must be increased. We found another LCP codes of dimensions **[89,45,17,17]** and a shortened code derive from it, i.e., **[86,42,17,17]** [20]. Both these codes will result in a better protection at the cost of chip area.

Then, we constructed another set of LCP codes with $d_{\text{Trigger}} \neq d_{\text{Payload}}$. In this time three different LCP codes with parameters **[73,37,17,12]**, **[73,37,17,8]** and **[49,37,5,3]** are applied. The process to apply the three codes is exactly the same.

The result of synthesis for the encoded nanoprocessor is presented in Table 8.1. This table shows the total gates, the area as well as the LCP code parameters as length, codeword, and security parameters d_{Trigger} and d_{Payload}. We can notice that the number of sequential gates increased from 37 to 73/86/89/81/49. It is logical because we encoded k flip-flops into n. The combinational logic part also increased due to the integration of $G, H, K,$ *and* J matrices. Pre-synthesis and post-synthesis simulations are performed to ensure that the encoded processor works correctly.

We can also notice that with the codes with different security parameters, we can reduce the overhead from 9717 to 7377/7324 μm^2 (new codes) with the same d_{Trigger} parameter. And the area is reduced from 7377 to 7324 μm^2 by reducing the d_{Payload} from 12 to 8 in the new codes. So with a smaller d_{Payload}, we can reduce the overhead of encoded method. For the last codes example (**[49,37,5,3]**) of the nanoprocessor, the overhead of this code is $< 3\times$. It could be acceptable for certain applications.

Table 8.2 Synthesis results of encoded circuit method, and security parameters for the SIMON coprocessor

IC (Code)	Gates	Area (μm^2)	n	k	$d_{Trigger}$	$d_{Payload}$
Original ([109,109,1])	300	1919	109	109	1	1
Encoded ([110,109,2,1])	560	3567	110	109	2	1
Encoded ([140,109,10,6])	3107	20239	140	109	10	6
Encoded ([123,109,5,3])	2348	15249	123	109	5	3

8.5.2 *Case Study II: SIMON Cryptography Coprocessor*

In the second case study, we use a lightweight crypto-processor SIMON. This cipherblock use a 32-bits plaintext and 64-bits of key and compute 32-bit ciphertext after 32 rounds. The unprotected SIMON gives the following after synthesis:

- 109 sequential cells (flip-flops),
- 300 combinational gates.

Thus we have $k = 109$ for the original SIMON netlist.

It is interesting to compare "encoded circuits" with "private circuits" [17]. Private circuits have an overhead quadratic with d_{Trigger}. For example, for $d_{\text{Trigger}} = 2$ (i.e., resistance to a single probe attack), the overhead is 39.7× in area (results obtained on Virtex5 FPGA). For the sake of comparison, we coded SIMON with LCP codes such that $d_{\text{Trigger}} = 2$ and $d_{\text{Payload}} = 1$. Table 8.2 shows that the overhead is only 1.9×, for the same level of security, using LCP.

In addition, Table 8.2 also presents the results of two LCP codes **[140,109,10,6]** and **[123,109,5,3]** on SIMON. This application shows that the encoding method can be used for very different ICs. These encoded SIMON circuits are also implemented on the FPGA to evaluate their performance against physical attacks. The analyses against physical attacks are presented in Sect. 8.7.

8.5.3 *Case Study III: AVR Processor*

In order to validate the automated insertion of encoding method, we use a larger circuit: an AVR processor. Specifically, we consider the AVR V14 downloaded from http://opencores.org/. This processor has 32×8 general-purpose registers and 23 interrupt vectors. In this case study, we repeat the same method as for the nanoprocessor. In the flat netlist of AVR, there are 468 sequential gates. Therefore, we constructed the LCP codes **[936,468,31,31]** for AVR processor. The results of the encoded AVR processor are also presented in Table 8.3

The encoder and decoder matrices G, H, and J matrices suitable in dimension for the AVR are represented in Fig. 8.7. The blue pixels are symbols for a 1, whereas the blank pixels stand for a 0. The synthesis results for those matrices are given in

Table 8.3 Synthesis results of AVR processor encoded circuit method, and security parameters

IC (Codes)	Gates	Area μm^2	n	k	$d_{Trigger}$	$d_{Payload}$
Original ([468,468,1,1])	2082	14015	468	468	1	1
Encoded ([937,468,31,31])	94969	586213	937	468	31	31

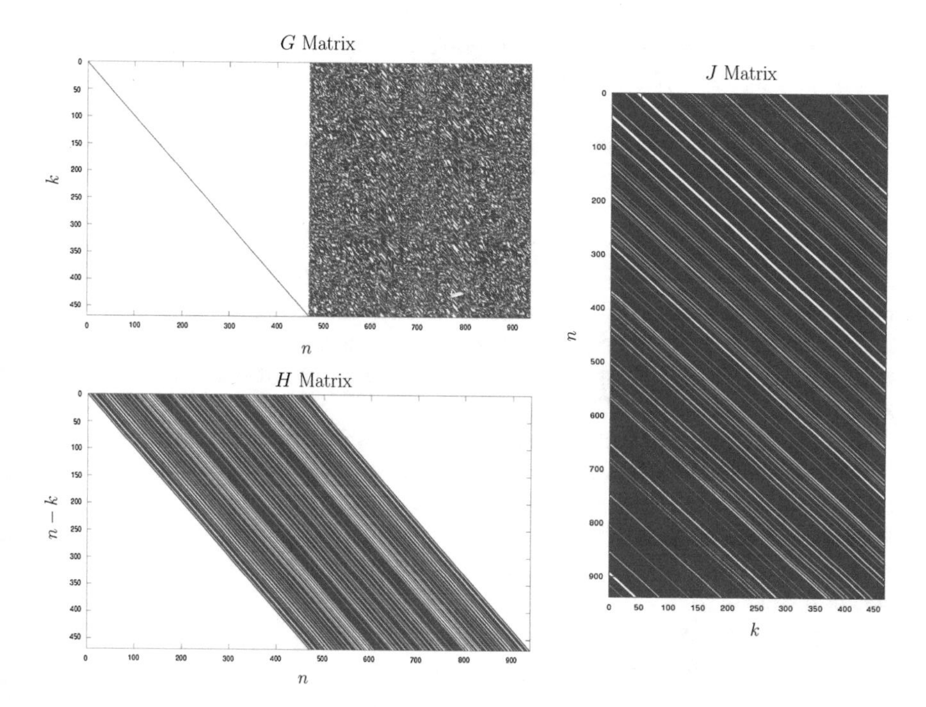

Fig. 8.7 Encoder and decoder matrices for the "encoded circuit" protection of the AVR core

Table 8.4 Synthesis results of G, H, and J matrixes for $k = 468$

Design	Number of		Logic gates (number)	Total area (μm^2)	Total power (nW)
	Rows	Columns			
G matrix	468	937	22811	137008	48059549
H matrix	469	937	26624	150248	56778482
J matrix	937	468	52546	302725	118381722

Table 8.4. We notice that the number of logic gates, power, and area increases significantly because of the complexity of G, H, and J matrixes. Certainly our protection can be optimized. Therefore, in the next section, we present the methodology used to optimize the LCP codes by keeping the same security parameters.

8.6 Optimization of LCP Code

8.6.1 Methodology

We aim at minimizing the hardware cost of the coding and decoding matrices, for a given pair of complementary codes C and D. It is very difficult to estimate the minimum number of gates required to synthesize the four applications of interest, namely $x \mapsto xG$, $y \mapsto yH$, $z \mapsto zJ$, and $z \mapsto zK$. A single-bit output, which consists in the application of a linear Boolean function, can be written very simply. For instance, let us denote by $G[i,j]$ the elements of matrix G. Then, let us denote by G_j, $1 \leq j \leq n$, the coordinate j of $x \mapsto xG$. It is equal to $G_j(x) = \bigoplus_{\substack{1 \leq i \leq k \\ G[i,j]=1}} x_i$, which requires a number of exclusive-or gates equal to the number of ones in the column of index j of G minus the number one. Thus, an upper bound on the number of exclusive-or gates required to synthesize the function $x \mapsto xG$ is the number of ones in G minus the number n. Of course, better results can be obtained by considering the coordinates not individually, but as a whole. Therefore, optimizations known as *common subexpression elimination* (CSE) can apply.

Still, unless the matrices are specially crafted, CSE will not improve a lot the size of the implementation. Thus, the number of ones in G is a good approximate indicator of the complexity to implement it in hardware. Consequently, we decide to quantify the cost of a matrix multiplication by this figure. In this respect, we introduce the notion of Hamming weight of a matrix. It is also sometimes referred to as the *grand sum* of the matrix.

Definition 6 The Hamming weight of a matrix G with elements in $\mathbb{F}_2$ is equal to its number of ones. It is denoted as $w_H(G)$.

In summary, our objective is to find a pair of complementary codes C and D, of generator matrices, respectively, G and H, such that conditions on distances are met, and such that

$$\mathrm{Cost}(G,H) = w_H\left(\begin{pmatrix} G \\ H \end{pmatrix}\right) + w_H\left(\begin{pmatrix} G \\ H \end{pmatrix}^{-1}\right) \tag{8.1}$$

is minimal.

Some literature exists on the topic of making matrices sparse [13]. However, in this section, we use an empirical method. Indeed, it is not obvious to assess the effect of working on matrices and their inverses simultaneously. As we shall see, our goal can be formulated as a *genetic algorithm*.

The idea is the following. We first find G and H such that the matrix $\begin{pmatrix} G \\ H \end{pmatrix}$ is invertible, based on some constraints. Refer for instance to [22, Sect. III]. Then, we randomly change the basis vectors of codes C and D spawned by generator matrices G and H, and test for a decrease in $\mathrm{Cost}(G,H)$.

To do so, we notice that any change of basis in G consists in linear combinations between the rows of G, which amounts to the left multiplication of G by an invertible matrix S. Indeed, owing to the invertibility of S, the matrix $k \times n$ matrix SG is also a maximal rank k. Therefore, a simultaneous basis change for C and D can be achieved by the following matrix multiplication:

$$\left(\frac{G'}{H'}\right) = \left(\begin{array}{c|c} S & 0 \\ \hline 0 & T \end{array}\right) \left(\frac{G}{H}\right) .$$

In this equation, the new generator matrix of C (resp. D) is G' (resp. H'). We insist that whatever the invertible matrices S and T, respectively, of size $k \times k$ and $(n - k) \times (n - k)$, $G' = SG$ and $H' = TH$ generate codes with identical parameters as C and D. When the choice for the new generating matrices G' and H' is done, so is $J' = JS^{-1}$ and $K' = KT^{-1}$, because

$$\left(J' | K' \right) = \left(\frac{G'}{H'}\right)^{-1} = \left(J | K \right) \left(\begin{array}{c|c} S^{-1} & 0 \\ \hline 0 & T^{-1} \end{array}\right) .$$

Now, exploring all possible basis changes through S and T is computationally infeasible. Moreover, application of random matrices S and T immediately turns the density into $\approx n^2$ (random matrix, with as many ones as zeroes). We have collected statistics based on 70,000 applications of random S and T. We have plotted them in Fig. 8.8; it can be seen that the grand sum starting from the situation where G and H are in systematic form (hence sparse). Clearly, although the initial value is 13160, the weight moves to value around $n^2 = 19600$ as soon as we multiply them by a random S or T.

Therefore, we explore the repetitive execution of a probabilistic algorithm, where the update of the generator bases is done:

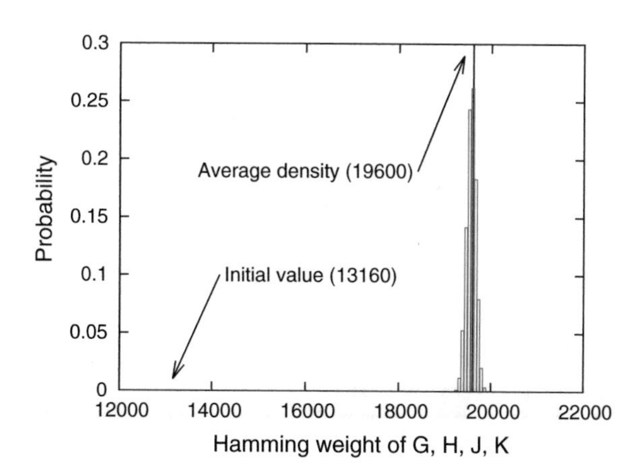

Fig. 8.8 Evolution of the grand sum when changing for a fully different base

- exclusively on G or H,
- by a simple linear combination consisting in adding two (different) base vectors to derive a new one.

The new bases are kept only if the total cost (Eq. (8.1)) is smaller. In this respect, this method is referred to as a *genetic algorithm*.

Algorithm 1: Genetic algorithmic to minimize the cost (Eq. (8.1)) of the pair of codes.

Input : $G = \left(I_k | M \right)$ and $H = \left(N | I_{n-k} \right)$ matrices, such that Proposition 3 is fulfilled.
Output: G' and H', spanning the same codes as G and H, but with smaller cost (Eq. (8.1)).

1 cost_ref $\leftarrow w_H\left(\left(\frac{G}{H}\right)\right) + w_H\left(\left(\frac{G}{H}\right)^{-1}\right)$
2 **while** *True* **do**
3 $\quad S \leftarrow I_k$
4 $\quad T \leftarrow I_{n-k}$
5 $\quad \text{coin} \xleftarrow{\$} \{0,1\}$ `/* Working either on` (G,J) `or on` (H,K) `*/`
6 $\quad$ **if** *coin* $= 1$ **then**
7 $\quad\quad i \xleftarrow{\$} \{1 \dots k\}$
8 $\quad\quad j \xleftarrow{\$} \{1 \dots k\} \backslash \{i\}$
9 $\quad\quad S[i,j] \leftarrow 1$
$\quad\quad$ `/* Left multiplication of` G `by` S `yields` G`, where` j`th row is added in-place to` i`th row */`
10 $\quad$ **else**
11 $\quad\quad i \xleftarrow{\$} \{1 \dots n-k\}$
12 $\quad\quad j \xleftarrow{\$} \{1 \dots n-k\} \backslash \{i\}$
13 $\quad\quad T[i,j] \leftarrow 1$
$\quad\quad$ `/* Left multiplication of` H `by` T `yields` H`, where` j`th row is added in-place to` i`th row */`
14 $\quad$ **end**
15 $\quad M \leftarrow \left(\begin{array}{c|c} S & 0 \\ \hline 0 & T \end{array}\right)\left(\frac{G}{H}\right)$
16 $\quad \text{cost\_new} \leftarrow w_H(M) + w_H\left(M^{-1}\right)$
17 $\quad$ **if** *cost_new* $<$ *cost_ref* `/* New record. Updating the state */`
18 $\quad$ **then**
19 $\quad\quad$ cost_ref $\leftarrow$ cost_new
20 $\quad\quad G \leftarrow SG$
21 $\quad\quad H \leftarrow TH$
$\quad\quad$ `/* Break the infinite loop if cost_ref is "`*good enough*`" */`
22 $\quad$ **end**
23 **end**
24 **return** $(G', H') = (G, H)$

8.6.2 Application on a LCP Pair of Codes

Algorithm 1 is applied on some codes. For instance, we use the example of the SIMON block cipher coprocessor described in Table II of [22]. It has parameters $[140, 109, 10, 6]$, meaning that, in the complementary pair C and D,

- the length of C and D is 140,
- the dimension of C and D is 109 and $140 - 109 = 31$,
- the dual distance of D is $d_{\text{Trigger}} = 10$, and
- the minimal distance of C is $d_{\text{Payload}} = 6$.

Before optimization, this code has cost (as per Eq. (8.1)) equal to 13160. After a couple of minutes, Algorithm 1 optimizes the cost to 11283, i.e., a reduction of 14.3 %.

Notice that the expected density of the (G, H, J, K) matrices output by Proposition 3 is

$$\frac{n + \frac{1}{2} \times k \times (n-k) \times 2 + \frac{1}{2} \times n^2}{2 \times n^2} = 34.0\,\% \ .$$

In reality the codes we have chosen have similar density, namely $\frac{13160}{2 \times n^2} = 33.6\,\%$. After calling Algorithm 1, the density is reduced to: $\frac{11283}{2 \times n^2} = 28.8\,\%$. After optimization, one could expect that the density of all matrices tend to the same value. Now, we observe that the Algorithm 1 stalls when J remains significantly of larger weight than other matrices. Recall that, in practice, we start Algorithm 1 with low-weight G and H matrices (they are in systematic form).

Therefore, we also tested two variants of Algorithm 1. First, we simply exchanged in Algorithm 1 the role of (G, H) and (J, K). More precisely, (G, H) are traded by $(J^{\mathsf{T}}, K^{\mathsf{T}})$. The initial convergence speed is larger, but the algorithm does not seem allow gaining more in terms of Hamming weight.

Second, we decide not to update J by a random addition of one row to another one, but by adding together the two rows which are of lowest Hamming distance. That is, lines 7 and 8 in Algorithm 1 are replaced by a exhaustive search of the nearest rows in J^{T} amongst the $\binom{k}{2}$ pairs of different rows. This strategy is working, as shown in Fig. 8.9. After 10 h of computations (on an Intel XEON running at 2.67 GHz), the Hamming weight is reduced from 13160 to 11223, i.e., by 14.7%. The quantitative comparison of the original Algorithm 1 and this improved version, in terms of Hamming weight and densities of matrices (G, H, J, K), is given in Table 8.5. Certainly some further improvements can be obtained, as the density of matrices is strongly unbalanced. This can be seen in Fig. 8.10, where the initial matrices are plotted on the left, and the new ones on the right. Clearly, there remains a strong dependency with the initial structure of the matrices.

Therefore, we test still another optimization. Basically, the three first strategies (whose performance is depicted in Fig. 8.9) stall after a given number of iterations. This means that a *local minimum* has been reached. Indeed, it is possible that whatever the pairwise addition of basis vectors, the grand sum does not decrease. For this

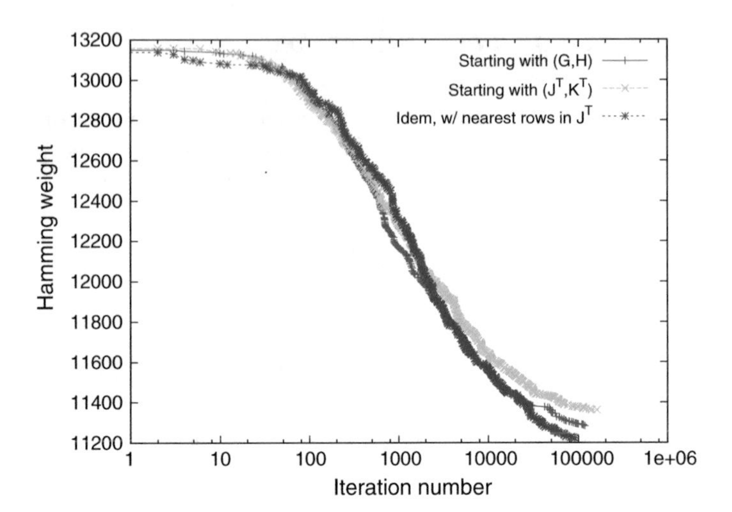

Fig. 8.9 Comparison between the Hamming weights of the quadruple of matrices G, H, J, and K, as a function of the iteration number at lines 2–23 of Algorithm 1, when starting with (G, H) and $(J^{\mathsf{T}}, K^{\mathsf{T}})$

Table 8.5 Hamming weight and densities per matrix for Algorithm 1 and its improvement

Matrix	Algorithm 1		Algorithm 1 improved	
	w_H	Density (%)	w_H	Density (%)
G (109 × 140)	2003	13.13	1930	12.65
H (31 × 140)	1493	34.40	1497	34.49
J (140 × 109)	5960	39.06	5944	38.95
K (140 × 31)	1827	42.10	1852	42.67

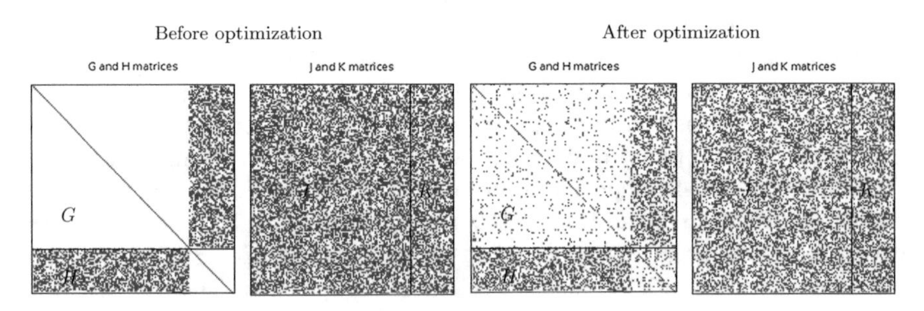

Fig. 8.10 Graphical representation of the (G, H) and (J, K) matrices, before and after optimization (also refer to Table 8.5)

reason, we allow to make more transformations to the basis. An arbitrary threshold of stall = 100 iterations is chosen. If after 100 iterations, there is no improvement, then we allow the matrices S or T to combine more than two vectors at once.

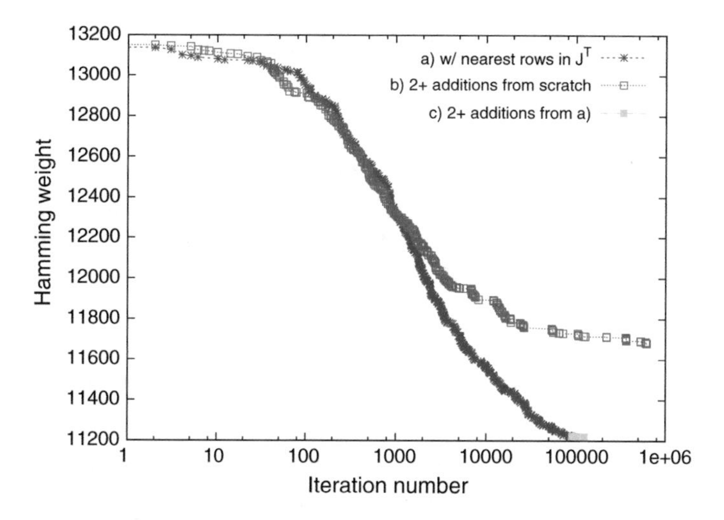

Fig. 8.11 Comparison between the Hamming weight of the quadruple of matrices G, H, J, and K, as a function of the iteration number, when one allows to combine more than two basis vectors

More precisely, S and T are still chosen initially to be identity matrices, but the number of nonzero extra-diagonal entries is allowed to be equal to 1, 2, 3, etc. The formula we choose is $1 + \lfloor \mathsf{stall}/100 \rfloor$. Of course, we check that S and T remain invertible, otherwise another random transform is selected.

The result is an improvement in *efficiency*, at the expense of a worse *efficacity*. Indeed, in Fig. 8.11, we compare

(a) the third strategy from Fig. 8.9,
(b) the new strategy where we allow more additions when the iterations stall,
(c) the same as (b), except that the threshold is set at $\mathsf{stall} = 1000$, and that it is only activated from the optimal solution (a) when it stalls.

The pure new strategy (in (b)) is very slow (loss of *efficacity*), actually slower than strategy (a). But the new strategy all the same allows to improve (a), as we can see in (c) (hence a gain of *efficiency*). Still, we notice that the gain in strategy (c) is small: with ≈12 h of computation, the grand sum is reduced from 11223 to 11220, which is very marginal.

8.7 Security Evaluation

In this section, we evaluate the encoded circuit method performance against other physical attacks as probing attack, side-channel attack, and fault injection attack.

8.7.1 Encoded Circuit Against Probing Attack

As encoded circuits are based on private circuits, they directly address the threat of **probing attack**. Probing attacks (front side or back side [14]) use tiny probes to monitor the inputs/outputs of internal blocks to directly recover sensitive data. By encoding all internal sequential logic parts and by masking encoded data with random numbers, this prevention method can also protect IC against probing attack with less than $d_{\text{Trigger}} - 1$ probes.

8.7.2 Encoded Circuit Against Side-Channel Attack (SCA)

SCA extracts sensitive information from a circuit using the power consumption, electromagnetic leakage, or delay analysis. We applied SCA on the encoded SIMON circuit (as in Table 8.2). The platform setup is a Sasebo GII FPGA Board (which contains a Virtex-5 FPGA) running at 24 MHz, *Langer RFU* 5 − 2 EM probe. For each design we acquired 200.000 traces with random plaintexts.

Next we use leakage detection techniques to check traces for any first-order leakage. We precisely compute the normalized inter-class variance (NICV) with respect to the plaintext in the traces. NICV is computed as $NICV = \frac{\mathsf{Var}[\mathbb{E}[T|X]]}{\mathsf{Var}[T]}$, where T denotes side-channel traces and X represents a chosen nibble of plaintext. Figure 8.12 shows four NICV computations for one-key byte of four different designs: uncoded SIMON, encoded SIMON [123,109,5,3] with RNG deactivated, encoded SIMON [123,109,5,3] with RNG activated and encoded SIMON [140,109,10,6] with RNG activated. By observing the results, we notice the following conclusions. First of all, encoded circuit without RNG should not be used, as the side-channel leakage is amplified. This is because during encoding we transform the code

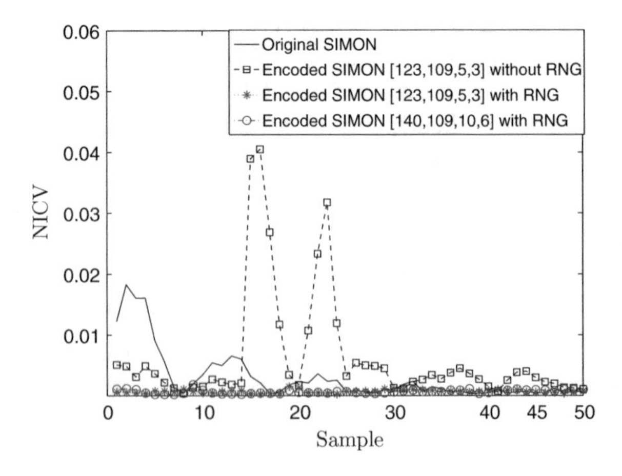

Fig. 8.12 NICV for encoded SIMON circuit

linearly without confusion, thus any linear distinguisher can detect the leakage. Second, encoded circuit with RNG do reduce side-channel leakage to an extent. The gain in SCA resistance is bounded due to limited entropy. Finally, as we increase d_{Trigger} from 5 to 10, we increase the entropy and therefore reduce the side-channel leakage.

We also compute the inverse security gain, i.e., the ratio between the NICV maximum values of encoded SIMON circuits with the one of original SIMON circuits. The inverse security gain w.r.t to original SIMON is 3.7 for encoded SIMON [123,109,5,3] without RNG, 0.0996 for encoded SIMON [123,109,5,3] with RNG, and 0.085 for SIMON [123,109,10,6] with RNG. The result shows that the signal-to-noise ratio (SNR) reduces significantly with the encoded circuit and d_{Trigger} can be used as the security parameter of side-channel attacks.

8.7.3 Encoded Circuit Against Fault Injection Attack

As stated earlier, encoded circuits can also be used to detect **Fault Injection Attacks** (FIA). This can be done by decoding and verifying the random numbers injected to mask the encoded circuit. Precisely, decoding can be done using the matrix K and the compared as shown in Fig. 8.5. If the input and output random differ, an "alarm" signal is raised and recovery mechanism like global reset is launched.

To evaluate this aspect of encoded circuit, we implemented the encoded SIMON [123,109,5,3] on the Sasebo-W FPGA board and UART for external communications. Then we perform global and local FIA on this board. The global FIA is done by varying the circuit frequency to inject the faults. 1000 tests were performed and "alarm" signal of encoded SIMON is activated 1000 times, i.e., a detection rate of 100 % Moreover, faults in individual rounds are detected separately.

For local FIA, we used electromagnetic injection (EMI). For the best evaluation of the encoded circuit against local FIA, we separate the location of each block on FPGA. The encoded SIMON is isolated from UART block and RNG block. Then we inject a single electromagnetic signal on the encoded SIMON. It insures that we will fault only encoded SIMON. We inject faults using an electromagnetic pulse of width of 1.5 ns. The EM pulse is injected from the beginning to the end of SIMON computation with a step of 1 ns. So in total, we perform 2560 steps. For each step, we perform 10 EMI. In each experiment, SIMON computes with the same plaintext and key. For 2560 delay steps EMI, there are statistically 2557 cases where the number of detection is equal to the number of faults. It means that we have a detection probability of 99.8 %.

So these first experiments demonstrate that encoded circuit method can detect FIA even with a high-cost FIA technique (EMI FIA).

8.8 Comparison with the State of the Art

8.8.1 Difference from Private Circuits

The proposed encoded circuit method prevents HT insertion at two different levels. First, like private circuits [17], it prevents the HT from retrieving any sensitive data by eavesdropping $< d$ flip-flops. This protection impedes the insertion of HT *trigger part*. Moreover, for the HT *payload part*, the proposed countermeasure also brings another aspect of active HT detection. If somehow the HT is able to write a malicious value into the state, the encoded state can be checked for errors introduced by the HT. For a HT to be functional, its payload must also be encoded with the same code as the original circuit (i.e., C and D matrixes). The case study on SIMON coprocessor shows that the hardware cost for encoded circuit with $d_{\text{Trigger}} = 2$ and d_{Payload} is 1.9× comparing to 39.7× of Private Circuit for the same security level [25]. So LCP method presents a big advantage comparing to Private Circuits in term of hardware implementation overhead.

8.8.2 Comparison with Previous Works

Preventing HT insertion by encoding internal variables of a circuit has been partially dealt in few previous works. Chakraborty et al. [9] initially presented a prevention method which obfuscates only the state machine of the IC. It is inspired by obfuscation methods [4, 10] initially intended to protect against IC counterfeiting. It partitions the states into an original state space and an isolation state space. The original state space can only be reached using a specific input pattern (e.g., secret key). If a wrong input pattern is presented at the input, the IC locks itself in a nonreversible isolation state space. Presented technique protects only the control part, while the data-sensitive part remains attackable. Instead, encoded circuits protect both parts (control and data). Moreover in [9] or even in [12], when the IC is well configured to reach the original state, it operates normally and cannot resist other physical attacks. Using encoded circuits, we can not only protect against HT insertion attack but also against others physical attacks because of the use of random numbers. The tests in SIMON cryptographic coprocessor confirmed this affirmation.

Another prevention method, ODETTE [5], is more intended to raise the HT activity for a better detectability than a proactive prevention. Furthermore, each bit of the state is masked with one bit of secret. With our method, we provide a more flexible solution, where the number of "mask" bits can be chosen, thus allowing the designer to adjust the security level. In [26], authors propose the method named "EPIC" which encodes the combinational logic part whereas in our encoded circuit method, the sequential logic part is encoded. EPIC is based on "security by obscurity" hence probing can be done after configuration to recover the key. This EPIC method is static; therefore, an attacker can create a HT which learns the key and subsequently gets activated, hence bypassing the EPIC method, whereas our "encoded circuit method" is dynamic because the circuit is encoded with a random mask, hence avoiding key learning attacks. Moreover, our method allows not only to prevent HT insertions but also to detect proactively HT insertions at runtime.

8.9 Conclusion

In this chapter, we proposed a provable randomization method, which encodes the IC using the linear complementary pair (LCP) codes C & D, allowing both HT detection and prevention.

It is based on quantifiable security parameters d_{Trigger} and d_{Payload} for HT insertion prevention and detection. Here d_{Trigger}, which is the dual distance of D, defines the minimum number of connections required to insert an effective HT. And d_{Payload}, which is the minimal distance of C, defines the minimum number of state that HT needs to modify, to be (hopefully) undetected. We studied the theory of codes and its rationale in "encoded circuits." Then we proposed the full automation CAD design for LCP method integration.

Many cases of studies are presented on SIMON cryptography coprocessor, nanoprocessor, as well as AVR processor. They demonstrated that LCP was successfully and automatically applied using Python and TCL scripts. These case studies also show that the hardware implementation cost will depend directly to the security parameters d_{Trigger} and d_{Payload}. By choosing small parameters, we can reduce significantly the overhead of LCP method but the security level will be decreased. So there is a trade-off between the implementation cost and security level.

Then, we talked about how to optimize the LCP methods while keeping the same security parameters d_{Trigger} and d_{Payload}. The algorithms in Chap. 6 shown that the hardware implementation cost can be significantly reduced. In this work, we used the LCP to encode all internal registers of circuit. But in real case, there could be only some parts which are needed to be protected against HT insertion. Therefore, we can apply the LCP only for these parts and not for the whole of circuit, hence reducing the hardware cost of our method.

Several tests on SIMON cryptographic coprocessor shown that LCP method is not only efficient against HT insertion but also robust against SCA and FIA. Using LCP method, we reduce significantly the SCA leakage by reducing the SNR. It also demonstrated that HT modifying less than d_{Payload} will be detected.

References

1. M. Abramovici, P. Bradley, Integrated circuit security: new threats and solutions, in *CSIIRW* (2009), p. 55
2. S. Adee, The hunt for the kill switch. IEEE Spectr. **45**(5), 34–39 (2008)

3. D. Agrawal, S. Baktir, D. Karakoyunlu, P. Rohatgi, B. Sunar, Trojan detection using IC fingerprinting, in *Proceedings of the 2007 IEEE Symposium on Security and Privacy*, SP '07, Washington, DC, USA. IEEE Computer Society (2007), pp. 296–310
4. Y.M. Alkabani, F. Koushanfar, Active hardware metering for intellectual property protection and security, in *Proceedings of 16th USENIX Security Symposium on USENIX Security Symposium*, SS'07, Berkeley, CA, USA. USENIX Association (2007), pp. 20:1–20:16
5. M. Banga, M.S. Hsiao, ODETTE: a non-scan design-for-test methodology for Trojan detection in ICs, in *International Workshop on Hardware-Oriented Security and Trust (HOST), IEEE* (2011), pp. 18–23
6. M. Banga, M.S. Hsiao, A novel sustained vector technique for the detection of hardware Trojans, in *Proceedings of the 2009 22nd International Conference on VLSI Design*, VLSID '09, Washington, DC, USA. IEEE Computer Society (2009), pp. 327–332
7. G. Bloom, B. Narahari, R. Simha, OS support for detecting Trojan circuit attacks, in *HOST* (2009), pp. 100–103
8. C. Carlet, S. Guilley, Complementary dual codes for counter-measures to side-channel attacks, in Springer, editor, *ICMCTA, 4th International Castle Meeting on Coding Theory and Applications*, CIM-MS, 15–18 Sept 2014. Palmela, Portugal, http://icmcta.web.ua.pt. (article #9). ISBN 978-3-319-17295-8. http://www.springer.com/978-3-319-17295-8
9. R.S. Chakraborty, S. Bhunia, Security against hardware Trojan through a novel application of design obfuscation, in *International Conference on Computer-Aided Design Digest of Technical Papers (ICCAD), IEEE* (2009), pp. 113–116
10. R.S. Chakraborty, S. Bhunia, Hardware protection and authentication through netlist level obfuscation, in *Proceedings of the 2008 IEEE/ACM International Conference on Computer-Aided Design*, ICCAD '08, Piscataway, NJ, USA. IEEE Press (2008), pp. 674–677
11. R.S. Chakraborty, F.G. Wolff, S. Paul, C.A. Papachristou, S. Bhunia, MERO: a statistical approach for hardware Trojan detection, in *CHES* (2009), pp. 396–410
12. S. Dupuis, P.-S. Ba, G. Di Natale, M.-L. Flottes, B. Rouzeyre, A novel hardware logic encryption technique for thwarting illegal overproduction and hardware Trojans, in *On-Line Testing Symposium (IOLTS), 2014 IEEE 20th International*, July 2014, pp. 49–54
13. L.-A. Gottlieb, T. Neylon, Matrix sparsification and the sparse null space problem. Algorithmica. **76**(2), 426–444. (2016) doi:10.1007/s00453-015-0042-6
14. C. Helfmeier, D. Nedospasov, C. Tarnovsky, J.S. Krissler, C. Boit, J.-P. Seifert, Breaking and entering through the silicon, in *ACM Conference on Computer and Communications Security*, ed. by A.-R. Sadeghi, V.D. Gligor, M. Yung (ACM, 2013), pp. 733–744
15. B. Hopkins, M. Beaumont, T. Newby, Hardware Trojans—prevention, detection, countermeasures, http://www.dtic.mil/cgi-bin/GetTRDoc?AD=ADA547668
16. Y. Ishai, M. Prabhakaran, A. Sahai, D. Wagner, Private circuits II: keeping secrets in tamperable circuits, in *EUROCRYPT. Lecture Notes in Computer Science*, vol. 4004. Springer, 28 May–1 June 2006. St. Petersburg, Russia, pp. 308–327
17. Y. Ishai, A. Sahai, D. Wagner, Private circuits: securing hardware against probing attacks, in *CRYPTO*. Lecture Notes in Computer Science, vol. 2729. Springer, 17–21 Aug 2003. Santa Barbara, California, USA, pp. 463–481
18. S. Jha, S.K. Jha, Randomization based probabilistic approach to detect trojan circuits, in *Proceedings of the 2008 11th IEEE High Assurance Systems Engineering Symposium*, HASE '08, Washington, DC, USA. IEEE Computer Society (2008), pp. 117–124
19. Y. Jin, Y. Makris, Hardware Trojan detection using path delay fingerprint, in *IEEE International Workshop on Hardware-Oriented Security and Trust, 2008. HOST 2008* (2008), pp. 51–57
20. J.L. Massey, Linear codes with complementary duals. Discrete Math. **106–107**, 337–342 (1992)
21. D.R. McIntyre, F.G. Wolff, C.A. Papachristou, S. Bhunia, Dynamic evaluation of hardware trust, in *Proceedings of the 2009 IEEE International Workshop on Hardware-Oriented Security and Trust*, HST '09, Washington, DC, USA. IEEE Computer Society (2009), pp. 108–111

22. X.T. Ngo, S. Bhasin, J.L. Danger, S. Guilley, Z. Najm, Linear complementary dual code improvement to strengthen encoded circuit against hardware Trojan horses, in *IEEE International Symposium on Hardware Oriented Security and Trust (HOST)* (2015), pp. 82–87. doi:10.1109/HST.2015.7140242
23. K. Nohl, E. Tews, R.-P. Weinmann, Cryptanalysis of the DECT standard cipher, in *FSE. Lecture Notes in Computer Science*, vol. 6147. Springer. Seoul, South Korea, 7–10 Feb 2010, pp. 1–18
24. R. Rad, J. Plusquellic, M. Tehranipoor, Sensitivity analysis to hardware Trojans using power supply transient signals, in *Proceedings of the 2008 IEEE International Workshop on Hardware-Oriented Security and Trust*, HST '08, Washington, DC, USA. IEEE Computer Society (2008), pp. 3–7
25. D.B. Roy, S. Bhasin, S. Guilley, J.-L. Danger, D. Mukhopadhyay, From theory to practice of private circuit: a cautionary note, in *The 33rd IEEE International Conference on Computer Design (ICCD '15)*, 18–21 Oct 2015, pp. 296–303. New York City, USA. doi:10.1109/ICCD.2015.7357117
26. J.A. Roy, F. Koushanfar, I.L. Markov, EPIC: ending piracy of integrated circuits, in *DATE* (2008), pp. 1069–1074
27. S. Skorobogatov, C. Woods, Breakthrough silicon scanning discovers backdoor in military chip, in *Proceedings of the 14th International Conference on Cryptographic Hardware and Embedded Systems*, CHES'12 (Springer, Berlin, 2012), pp. 23–40
28. M. Tehranipoor, F. Koushanfar, A survey of hardware Trojan taxonomy and detection. IEEE Des. Test **27**(1), 10–25 (2010)
29. M. Tehranipoor, C. Wang (eds.), *Introduction to Hardware Security and Trust* (Springer, 2012). ISBN 978-1-4419-8079-3
30. U.S. Department Of Defense: Defense science board task force on high performance microchip supply. http://www.acq.osd.mil/dsb/reports/2005-02-HPMS_Report_Final.pdf
31. X. Wang, M. Tehranipoor, J. Plusquellic, Detecting malicious inclusions in secure hardware: challenges and solutions, in *Proceedings of the 2008 IEEE International Workshop on Hardware-Oriented Security and Trust*, HST '08, Washington, DC, USA. IEEE Computer Society (2008), pp. 15–19
32. M.J. Wirthlin, B.L. Hutchings, K.L. Gilson, The Nano processor: a low resource reconfigurable processor, in *IEEE Workshop on FPGAs for Custom Computing Machines, 1994. Proceedings*, Apr 1994, pp. 23–30

Chapter 9
Ultra-Lightweight Implementation in Area of Block Ciphers

Cédric Marchand, Lilian Bossuet and Kris Gaj

9.1 Introduction

The research field of lightweight implementation of cryptographic algorithms is relatively new. It appears with the emergence of highly constrained applications such as radio-frequency identification (known as RFID) and wireless sensor networks. All these new applications require security and implement cryptographic algorithms in such constrained devices, which is challenging. The constraints that are involved in the implementation strategy strongly depend on the target application. In the past, lightweight implementations of cryptographic algorithms aim to be energy efficient because devices used for the target application do not contain any power source or limited one. In addition, these devices (RFID tag) are extremely small and the area of implementations is one of the main constraints. Today, numerous applications requiring lightweight hardware implementation of cryptographic functions have appeared. Among all primitives adapted for lightweight applications, the most common construction used is called block ciphers. This category of primitives is used to encrypt or decrypt only a block of data at a time. They are built with small functions organized in a round, which is applied multiple times. The most known block cipher adapted for lightweight applications is PRESENT [6] but many others have been published recently. Some of them are briefly described on the CryptoLux website [1].

In this jungle of lightweight implementation of block ciphers, comparing the results between different works is an absolute necessity but this task is extremely challenging due to the differences in the target applications. Indeed, comparing two works about lightweight implementations is possible but usually unfair because one application might not have the same constraints than the other. In addition, the com-

C. Marchand (✉) · L. Bossuet
Laboratoire Hubert Curien, University of Lyon, Saint-etienne, France
e-mail: cedric.marchand@univ-st-etienne.fr

L. Bossuet
e-mail: lilian.bossuet@univ-st-etienne.fr

K. Gaj
Department of ECE, Volgenau Schoole of IT&E, George Mason University, Fairfax, VA, USA
e-mail: kgaj@gmu.edu

L. Bossuet and L. Torres (eds.), *Foundations of Hardware IP Protection*,
DOI 10.1007/978-3-319-50380-6_9

parison may be also unfair if the interfaces used in two different works are not identical. The best way to compare hardware implementations fairly is thus to design a generic framework and use it to implement all the algorithms. If the comparison with other studies has to be done, it is mandatory to implement the algorithms using the exact same devices and carefully explain the differences in the target application constraints.

In this chapter, four different block ciphers (KLEIN, LED, Lilliput, and Ktantan) are presented and implemented using Xilinx Spartan 6 FPGA (45 nm CMOS technology) and Xilinx Spartan 3 (90 nm CMOS technology). Hardware implementations presented in this chapter are meant to be used in the SALWARE project [7]. In this project, the implemented algorithm will be used only once or very few times in the life time of the device and no one wants to pay for a large implementation of a block cipher that is used only one time. That is why the main constraint in this work is the area of the implemented algorithms. The block cipher that uses the minimum area will be chosen as the candidate for the SALWARE project. Additionally, this project does not target a specific technology and the use of FPGA aims to prototype ultra-small cryptographic algorithms to implement them using other FPGA technologies and ASIC. So, no specific feature of the used FPGA will be used. For example, Xilinx Spartan 6 FPGAs contains distributive RAM and specific shift registers that can be very interesting to reduce the area of an implementation but it is mandatory to not use them because of the portability requirement.

The comparison between the four implemented cryptographic algorithms is made using a generic framework that can be reuse in other studies due its strong adaptability to any other algorithms. Finally, to be able to understand the mechanisms of lightweight implementation of block ciphers, each algorithm is implemented using two different strategies. The first one is a fullwidth implementation which uses the block length for the width of the datapath. The second is a full serial implementation which uses the smallest possible word size for the width of the datapath. All results presented in this chapter contain both encryption and decryption.

The rest of this chapter is organized as follows: Sect. 9.2 gives general background about block ciphers algorithms for lightweight applications. The two implementation strategies used in this work are presented in Sect. 9.3. Section 9.4 describes how the four implemented algorithms have been chosen, gives an overview, and presents the two implementations of each block cipher. The results of these implementations are compared and discussed in Sect. 9.5. Finally, the last section concludes and gives some future interesting and possible development for this study.

9.2 Lightweight Block Ciphers

9.2.1 General Description

A block cipher is a symmetric-key algorithm used to encrypt or decrypt a block of data with a fixed length. To encrypt a block of data, a message (or plaintext) is transformed using a key in a ciphertext of the same length. An inverse function is used to decrypt the ciphertext using the same key. A block cipher is often composed of small functions organized to form a round that is performed several times. Additionally, the key is expanded, using a function called keyschedule, in round keys that are used during the rounds of the algorithm. Figure 9.1 shows how a block cipher can be seen with a general schematic.

The most known block ciphers are the advanced encryption standard (AES) [9] and the data encryption standard (DES) [22]. These two block ciphers are constructed with two different structures that are almost all the times reused for all other block ciphers: Substitution and permutation network for AES and Feistel network for DES. Recently, a lot of new block ciphers have been developed especially for lightweight applications and are commonly referenced under lightweight cryptography. PRESENT [6] is one of the first block ciphers designed for lightweight applications and commonly serves as reference to compare all others algorithms implemented for lightweight. The number of cryptographic primitives developed for this relatively new application has risen and some surveys ([11, 17]) try to compare many of them. In the most recent one [17], 50 block ciphers are compared using hardware, software, or both. Additionally, when a new family of block ciphers designed for lightweight applications is presented, comparison with previous work is almost always done in order to show the added value of the new family like in [4] where the two described families (SIMON and SPECK) are compared to five other cryptographic algorithms: Twine [23], Present [6], Piccolo [19], Katan [10], and KLEIN [12].

However, even if all these new algorithms are categorized under lightweight cryptography, this category does not really exist and it is more accurate to talk about lightweight implementation of block ciphers instead. Indeed, some regular algorithms such as the AES are also implemented for lightweight applications like in [8] or in [13]. Nevertheless, there are cryptographic algorithms that are adapted for lightweight applications. They usually use a smaller block size of data such as 32, 48, or 64 bits; the key size is also commonly reduced to 64, 80, or 96 bits even if some algorithms are able to use 128-bit key as well. The most common sizes used in

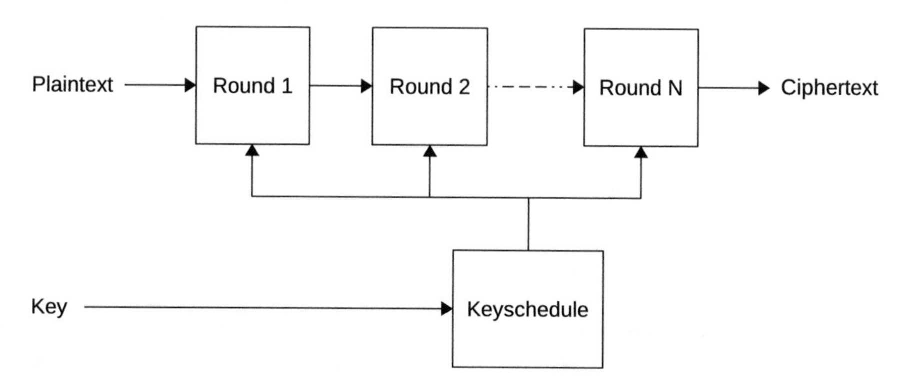

Fig. 9.1 General schematic of a block cipher

block ciphers for lightweight applications are 64-bit block and 80-bit key. Additionally, block ciphers that aim to be used for lightweight applications often operate on smaller word than regular block ciphers and use bit, or nibble (4-bit word) as word instead of bytes. Accordingly, all functions are smaller in size and build in order to be able to implement them using very few resources in FPGAs or ASICs. Finally, the keyschedule is also smaller and in some algorithms, this function only selects bits of the secret key without any logical transformations.

The two most known structures used for block ciphers are the substitution and permutation network (AES like) and the Feistel network (DES like). The two next subsections present the particularities of each one of these two constructions.

9.2.2 Substitution and Permutation Network

The most common construction of block ciphers is a substitution and permutation network (SPN) which is the construction used by the AES. This kind of block ciphers is usually built around three different layers that operate on the full block of data:

- The *add round key* layer is usually used to mixed the round key with the current state of data. It is commonly a bitwise XOR applied between the state of data and the corresponding round key.
- The *substitution* layer is often nonlinear and comprise substitution boxes called sboxes, which take a n-bit word in input and return a m-bit word at the output. Most of the block ciphers use 8-bit sbox that map a 8-bit word to another 8-bit word. However, in bloc ciphers designed for lightweight applications, sboxes are commonly nibble wise.
- The *diffusion* layer is most of the time a permutation function that can takes different sizes in input such as bit, nibble, or byte for example. This layer is often the bottleneck of any optimization because of the difficulty to reduce its area.

Due to the structure of this construction, the number of round used to encrypt or decrypt a block of data is smaller than for others construction such as Feistel network. They are easy to design in hardware and offer a good compromise between the area and the throughput of the implementation.

9.2.3 Feistel Network

Another very common construction of block ciphers is based on Feistel network like the DES. This construction also contains a round function based on the same three layers as the SPN structure but instead of applying it to the full state of data, this function changes only half of the data. Then, the permutation ensures to switch the two halves of the state in order to apply the round to the second part of the state during the next round. This structure is very famous for lightweight applications due

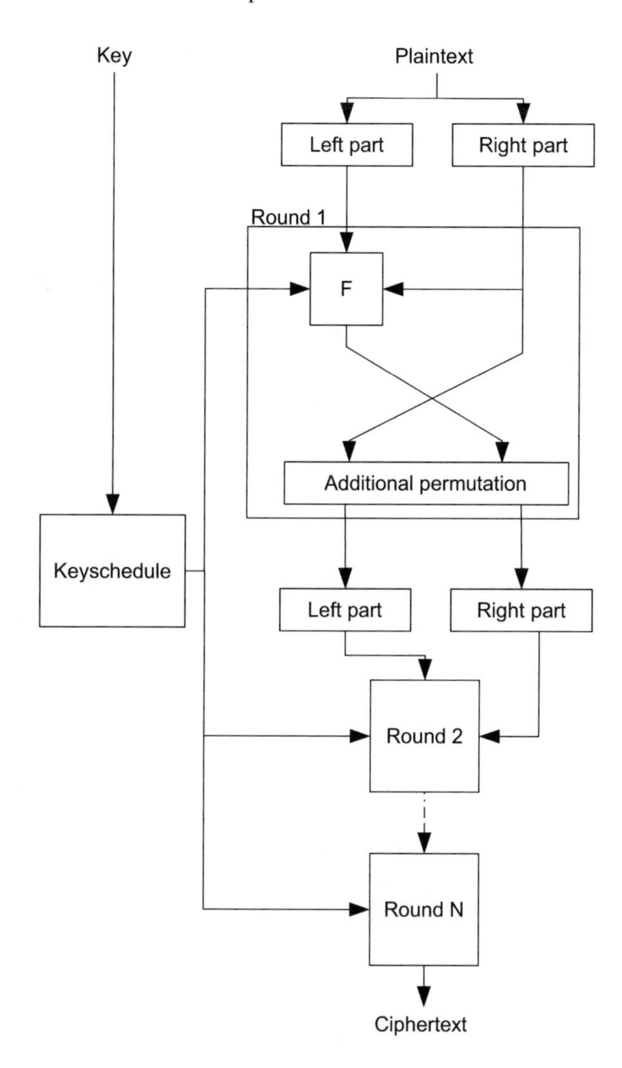

Fig. 9.2 General schematic of Feistel network

the keyschedule that is usually very simple. Furthermore, classical Feistel networks use exactly the same process for encryption and decryption which is very interesting for lightweight applications because only one encryption core can be implemented. Among block ciphers designed for lightweight that use this structure, it is possible to cite LBlock [24], ITUbee [16], or SEA [21] for example.

However, the permutation can also be more complicated than just switch the two halves and also performs a permutation inside the halves themselves. In this case, the structure of the block cipher is called generalized Feistel network (GFN) like Lilliput [5], Hight [15], or Clefia [20]. In this case, the inverse permutation has to be implemented to make the decryption possible. Figure 9.2 shows a general schematic that describes regular Feistel network and GFN both.

To describe classical Feistel network using Fig. 9.2, the additional permutation block is simply removed.

9.2.4 Other Types of Block Ciphers

One family of algorithms uses a construction which is completely different than the two previous. It is the Katan and Ktantan family of block ciphers [10]. This family is based on feedback shift registers updating only one to three bits at a time. Due to this, the construction is very similar to stream ciphers but these algorithms can be used only to encrypt or decrypt block of data and cannot be categorized as stream cipher. The number of round that is needed to achieve sufficient security is 254 which is largely bigger than for the two other construction of block ciphers.

9.3 Hardware Implementation Strategies

In this chapter, hardware implementations for both ASICs and FPGAs are discussed. At the end of the chapter, all the results are generated using Xilinx Spartan 6 FPGA and Xilinx Spartan 3 FPGA (which does not have the same logical elements) for prototyping.

Two main strategies are commonly used to implement block ciphers: fullwidth and serial implementations. These two different implementation strategies are presented in this section. Other strategies may be used depending on the goal of the implementation, but all are somewhere between the two and are commonly referenced as partial serial implementations [4].

9.3.1 Fullwidth Hardware Implementation

Fullwidth hardware implementation is probably the most simple strategy that can be used in order to implement a block cipher. This kind of implementations usually implements the basic round function of the algorithm once and applies it to the block of data several times. Each time, the full block is updated and the number of clock cycles required to encrypt one block is exactly the number of rounds needed to complete the cryptographic algorithm. If one layer of the round operates on a small word instead of the full state, this function is implemented multiple times in parallel in order to apply it to the full block of data in one clock cycle. In the exact same way as for the block of data, the key is transformed by the keyschedule is implemented only once in the device. If necessary, a subkey is generated before being sent to the round function and mixed with the state of data.

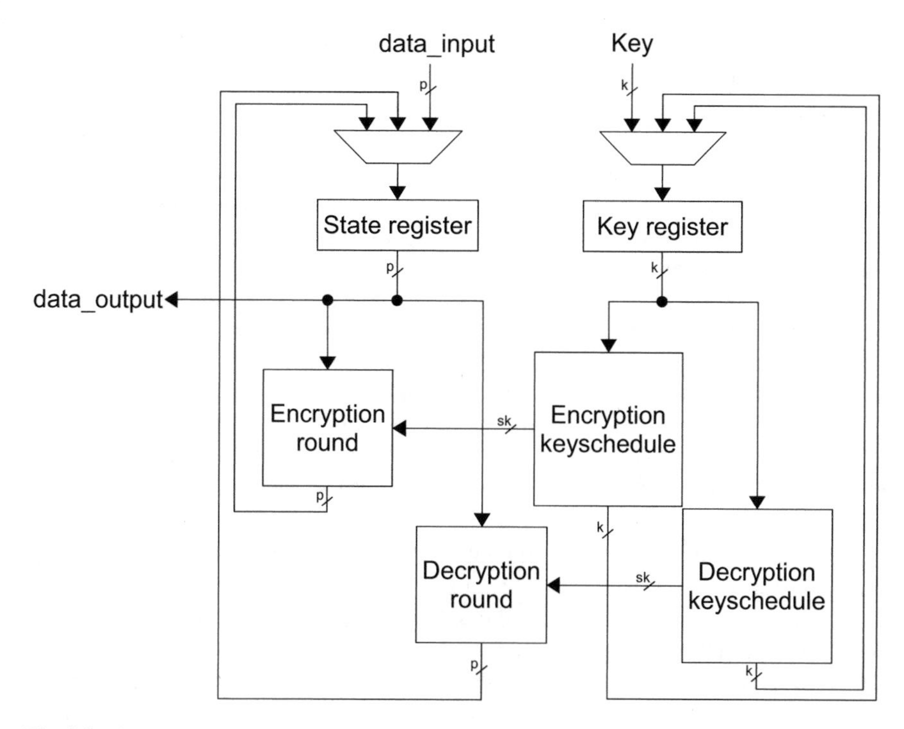

Fig. 9.3 General diagram of a fullwidth hardware implementation of a block cipher

This implementation strategy should be the starting point of every work that aims to reduce the area of the implementation since it is the most naive and simple way to implement one primitive. In addition, this strategy helps the designer clearly understand the functions involved in the round. Figure 9.3 presents a general diagram of both encryption and decryption that can be used for every fullwidth implementations with very small changes depending on the implemented algorithm. In some algorithms, there is almost no keyschedule and in others, part of the round can be shared between encryption and decryption. In these cases, Fig. 9.3 would simply needs some modifications to highlight these parts like in Sect. 9.4.2.1 or Sect. 9.4.3.1.

In Fig. 9.3, the plaintext has a length of p which is the size of the block to process, k is the size of the key, and sk is the size of the round keys. At the end of the computation of the ciphertext, the result is extracted directly from the state register.

9.3.2 *Serial Hardware Implementation*

This strategy is commonly used to implement block ciphers for lightweight applications. To successfully implement a block cipher using the serial strategy, each function needs to be analyzed to find the smallest possible word size that can be used.

A serial hardware implementation tries to use the minimum amount of resources in the devices using the smallest possible datapath. For this strategy, the execution time of the block cipher is sacrificed to the advantage of the area. For example, let us assume an algorithm is designed for block of data comprising 64 bits and that the used substitution box maps a nibble to another nibble. Instead of implementing 16 sboxes in parallel and processing the full block of data at one go, only one sbox is implemented for the serial implementation. Thus, to process the full block of data, the implemented sbox is used 16 times.

This is how serial implementation works; only the minimum number of functions is implemented and used several times to ensure the correct functionality. To do so, the state and the key registers are usually transformed into shift registers and only a small part of them is updated using the reduced round. However, in most cases, the keyschedule is very difficult to optimize and impossible for some block ciphers. In this case, the block of data is processed serially and the key is kept during multiple clock cycles to ensure that all the state of data has been updated using the correct round key.

However, the full serial hardware implementation strategy is not always the smallest implementation due to choice and storage penalties that may appear if one function requires more or fewer bits that the previous one. These penalties include additional multiplexers or registers and can increase the size of the finite state machine of the cryptographic core. Nevertheless, this strategy is commonly used as a reference to prove that a new algorithm is well suited for lightweight applications and requires only a small area.

The two strategies presented in this section are used in the rest of this chapter. All block ciphers presented in Sect. 9.4 have been implemented following the fullwidth and the full serial strategies. The results of hardware implementations are compared in Sect. 9.5.

9.4 Description of the Implemented Algorithms

9.4.1 Choice of the Algorithms to Implement

Before presenting the four block ciphers implemented in this study, let us explain how the algorithms have been chosen. At the beginning of the project, eight algorithms were preselected among those presented in the CryptoLux website based on their specifications, their publication dates, and the availability of test vectors: Hight, Lblock, Itubee, Lilliput, Present, LED, KLEIN, and Ktantan. Then, all these algorithms have been implemented using Matlab/Octave to understand them and to be able to easily generate new test vectors if needed. Additionally, using Octave to implement these block ciphers also facilitate debugging hardware implementations because it makes it very simple to show intermediate values which is not the case with C implementations. All these Matlab/Octave implementations are available on the SALWARE website [2].

The four algorithms have been chosen by looking at the size of the block to be processed as well as at the size of the key to use. According to the literature on the lightweight implementation of block ciphers, the use of a 64-bit block and 80-bit key is appropriate. Nevertheless, LED has been kept even if the key is 128 bits because of its very interesting diffusion layer. Table 9.1 gives basic information on the four selected algorithms: date of publication, references for specifications, the type of cipher, the size of the block, and the size of the key used in this chapter.

At the end, this chapter compares one Feistel network, one SPN, and one other construction using the exact same key and block length and one SPN that use a 128-bit key is also implemented and compared to the other block ciphers.

9.4.2 KLEIN

KLEIN is a block cipher taking a 64-bit block of data in input and can be used with three key sizes: 64, 80, or 96 bit. It has a SPN construction and the keyschedule is a variant of a Feistel network. The specifications of this algorithm have been published in 2012 [12].

To encrypt or decrypt a message of data, the number of round that needs to be performed depends on the key size and is respectively 12, 16, or 20 rounds for 64-, 80-, and 96-bit key. Each round is composed of the four classical functions used in others SPN-based cryptographic algorithms:

- *AddroundKey* is the first function and mixes the round key with the state of data using a bitwise xor.
- *SubNibbles* is the substitution step. It uses an involutive 4-bit sbox. Let us note S the sbox and x the nibble to process, and the involution property is defined as follows:

$$S(S(x)) = x$$

  Thus, the same sbox is used for both encryption and decryption which is very interesting to reduce the area of the implementation of this cipher.

Table 9.1 Bloc ciphers implemented in this chapter

Algorithm	Year	Reference	Structure	Key size(s)	Block size(s)
Ktantan	2009	[10]	Bloc/Stream	80	64
LED	2011	[14]	SPN	128	64
KLEIN	2012	[12]	SPN	80	64
Lilliput	2015	[5]	GFN	80	64

- *RotateNibbles* is the first function of the diffusion layer and performs a rotation of two nibbles per round
- *MixNibbles* is exactly the same function as the mix column of the AES algorithm and completes the diffusion layer.

In this chapter, only the 80-bit key version of the cipher is implemented. One advantage of this block cipher is the substitution box that is involutive and nibble wise. This is a real strength since the same sbox is used for encryption and decryption processes. KLEIN uses the AES mix column operation; the minimum input size for this layer is 32-bit and the minimum number of output is 8-bit. This choice of mix operation makes very difficult to implement KLEIN using a datapath smaller than 8-bit. Additionally, the rotation function is 8-bit wise and operates on the full state which also implies that the minimum datapath is 8-bit.

9.4.2.1 Fullwidth Implementation

For the fullwidth implementation, the sboxes used during the round are shared between encryption and decryption because of the involution property. Thus, the round is split into two parts in order to be able to share the SubNibbles function. Depending on the operation processed (encryption or decryption), the correct input is selected and the output of this layer is sent to the second part of the round for both encryption and decryption. Finally, the state is updated using the correct round output depending on the operation again. For the encryption process, the first part of the round only contains the AddroundKey and the second part is composed of the rotation and of the mix operation. On the other side, the first part of the decryption round contains the mix operation followed by the rotation and the second part only computes the AddroundKey function.

The keyschedule which is a variant of a Feistel network ensures that the right round key is correctly mixed with the state at the right time. However, due to its structure, the last round key has to be computed before starting any decryption operation that includes an additional latency of 16 clock cycles per decryption.

Figure 9.4 shows the datapath of this fullwidth implementation of the KLEIN core. Both encryption and decryption are represented.

9.4.2.2 Serial Implementation

To achieve a serial hardware implementation of KLEIN encryption, the state register is replaced by a shift register with multiple choices at the top of it. Only one byte is process per clock cycle for the two first functions. The rotation is included in the choice of the byte to be processed and in the multiplexers on top of the state register. For the mix operation, 32 bits are required to produce one byte at the output. Thus, an additional shift register is implemented to wait during the first four clock cycles to be able to update four bytes during the four following clock cycles. The mix operation

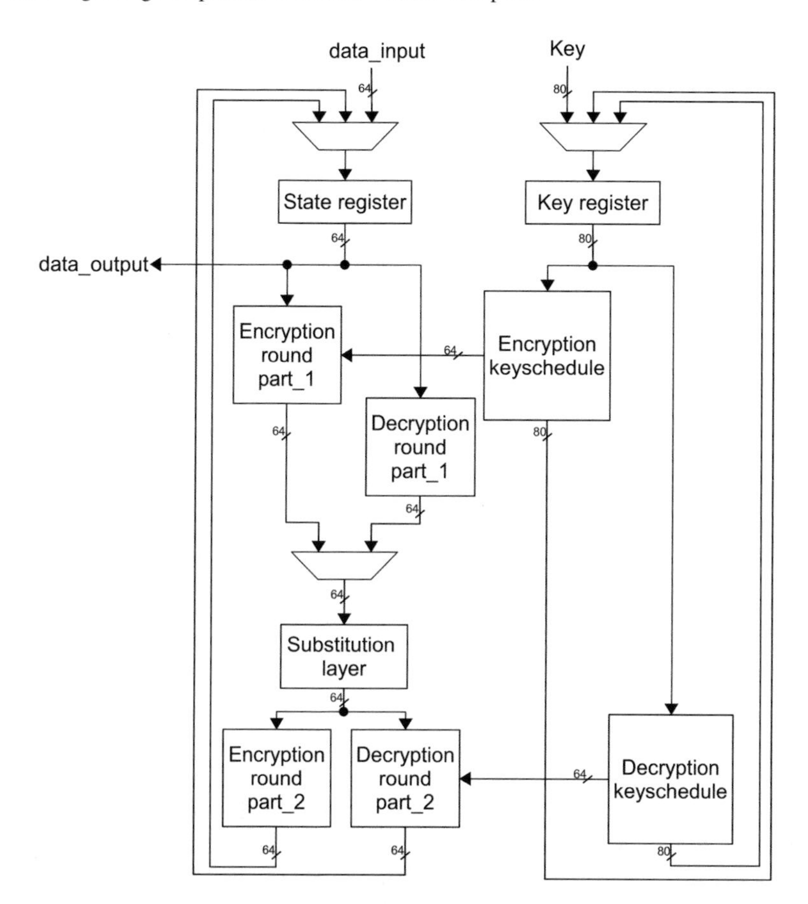

Fig. 9.4 Fullwidth datapath of KLEIN

itself is optimized to produce only one byte at a time instead of four. This process is applied twice in order to update the full state.

For the decryption process, a similar structure is used which includes different choices of the byte to update each clock cycle. In this case, the first operation to compute is the inverse mix function so the four first clock cycles are dedicated to the mix operation and the four following to the second part of the round. The process is also applied twice.

It is not possible to optimize the keyschedule because of its structure, so this operation is exactly the same as in the fullwidth implementation. Nevertheless, the correct key is kept and shifted by one byte each clock cycle to provide the correct byte of the round key at each clock cycle.

To make Fig. 9.5 more readable, only the encryption datapath of the serial implementation of KLEIN is shown. All the connections are 8 bits except if another size is noted.

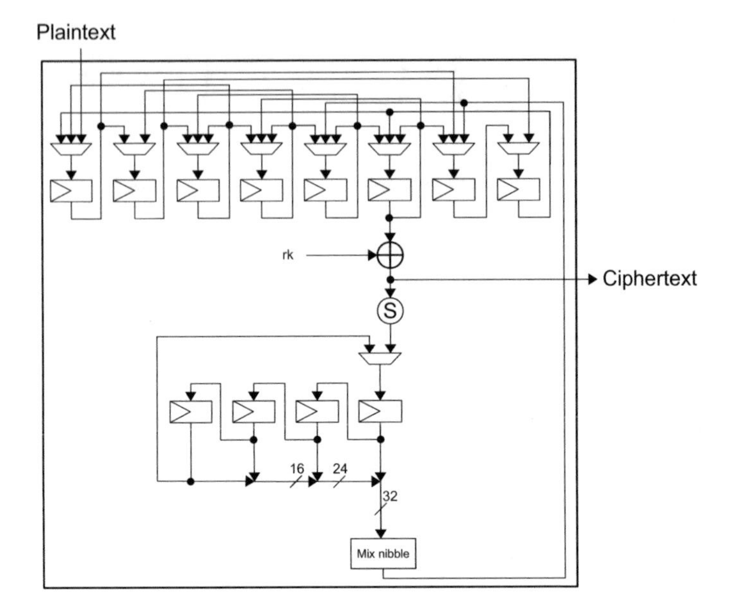

Fig. 9.5 8-bit optimized datapath of KLEIN encryption

9.4.3 LED

LED is the second algorithm implemented in this study. It is also a SPN but the key is mixed with the state of data each four rounds. The block size is 64-bit and the key can be of two different sizes: 64 or 128 bit. This block cipher is organized by step that contains four rounds. To process data for encryption or decryption, the number of steps is 8- for 64-bit key and 12- for 128-bit key. This block cipher has been published in 2011 [14]. In this work, only the 128-bit key version is implemented, so the number of step to be performed is 12 and the total number of round is 48. A general schematic for encryption is presented in Fig. 9.6.

As can be seen in Fig. 9.6, the LED block cipher does not contain a keyschedule. For the 128-bit key version, the round key used is alternatively the 64 less significant bits of the key and the 64 most significant bits. Additionally, due to number of step, the order of the part of the key that has to be applied is the same for encryption and decryption process and no latency is added for the decryption process.

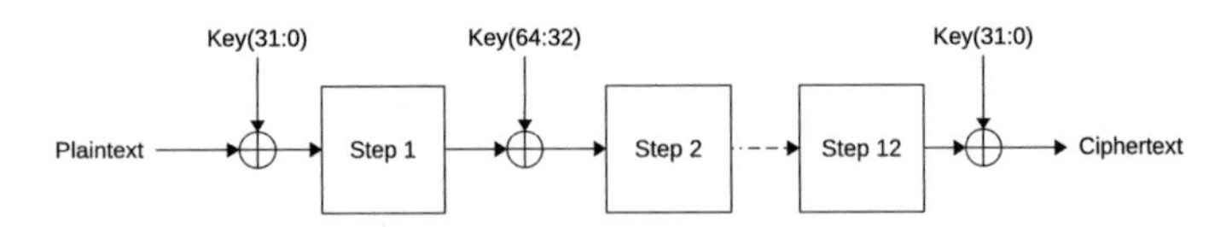

Fig. 9.6 General construction for the 128-bit version of LED

LED's round is composed of four different functions, namely AddConstants, SubCells, ShiftRows, and MixColumnSerial.

- The *AddConstants* operation is a bitwise xor which add a value to the half of the state of data.
- The *SubCells* is the substitution step that uses a 4-bit to 4-bit sbox. This sbox is not involutive as for KLEIN and the inverse sbox needs to be implemented for decryption.
- *Shiftrows* is the same operation as in the AES block cipher but operates on nibble instead of bytes.
- The *MixColumnSerial* is the real strength of this block cipher from implementation point of view. Indeed, it corresponds to a multiplication of the state by a matrix but the final operation can be reduced by using a very simple matrix that updates only one nibble at a time. The inverse operation also has a reduced matrix.

9.4.3.1 Fullwidth Implementation

For the fullwidth implementation of LED, keyschedule and the addition of the key with the state are shared between encryption and decryption. All the other functions have to be implemented for the two different operations.

In the encryption round, all four functions are implemented. For the Substitution layer, the 4-bit sbox is implemented 16 times to apply the function to the full state. The Shiftrows operation is exactly the same as in the AES algorithms except that it is apply to nibble instead of byte. The particularity of LED is the mix operation that uses a reduced matrix to update 16 bits. However, this reduced matrix is implemented four times in parallel to be able to process the full block of data in this fullwidth implementation of the cipher.

For decryption, the inverse mix column is also implemented four times using the reduced inverse matrix. The substitution layer contains 16 times the inverse sbox like in the encryption process. Figure 9.7 shows the datapath of the fullwidth implementation of this bloc cipher for both encryption and decryption.

9.4.3.2 Serial Implementation

The strength of the LED block cipher is that the mix operation can be applied serially to update only one nibble per clock cycle. The only function that requires the full state to be processed is the shift operation but it can be included in the multiplexers on top of the state register. Thus, for the serial implementation of LED, a 4-bit datapath is chosen. The round function takes 32 clock cycles, the first 16 are dedicated to the first three functions of the round and the 16 next to the mix column operation. For the decryption process, the first 16 are used to compute the inverse mix column and the last 16 are dedicated to the other functions of the round. Figure 9.8 presents only the encryption datapath for the serial implementation.

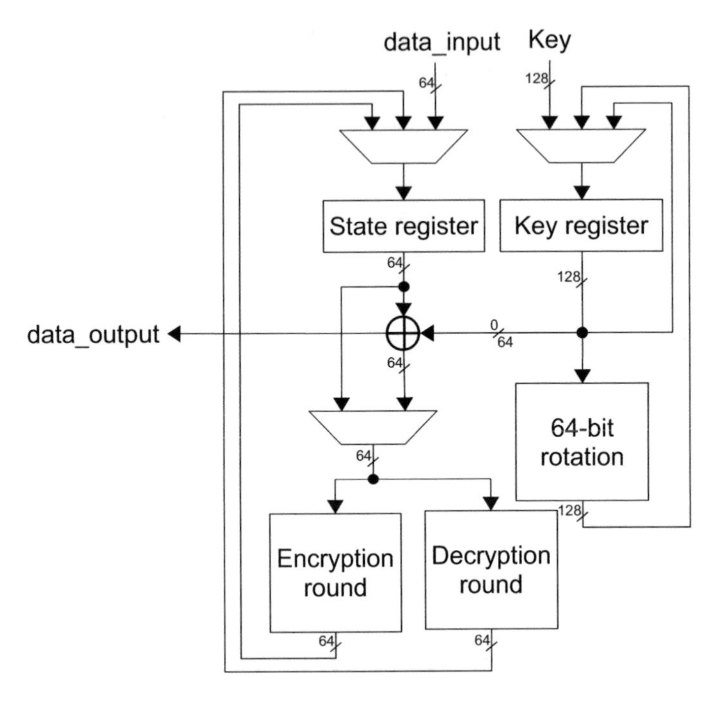

Fig. 9.7 Fullwidth datapath of LED

In Fig. 9.8, all the connections are 4-bit wide except for those whose width is indicated. The state register is replaced by a shift register with additional destinations for 12 nibble because of the shift operation. The latter function is performed during the 16th clock cycle of each round. The entire implementation is arranged to operate using a 4-bit datapath. Only the Mix operation represented by the matrix A takes 16 bits in input. However, even this function computes only a nibble and does not include a penalty to fall the datapath to 4 bits. Thus, to complete one round of the LED cipher using this serial implementation, 32 clock cycles are needed.

Finally, all parts of LED are reduced for the serial implementation except the multiplexers on top of the state register. This is the bottleneck of the implementation because it might lead to an increase in area. However, all the other functions are reduced and the total area of this 4-bit optimized datapath is expected to be small.

9.4.4 *Lilliput*

Lilliput is a block cipher that uses a generalized Feistel network structure. The block size is 64 bits and the key size is 80 bits. It has been published in 2015 [5].

The strength of the Feistel construction is that the same flow is used for encryption and decryption. Nevertheless, because there is an additional permutation at the end

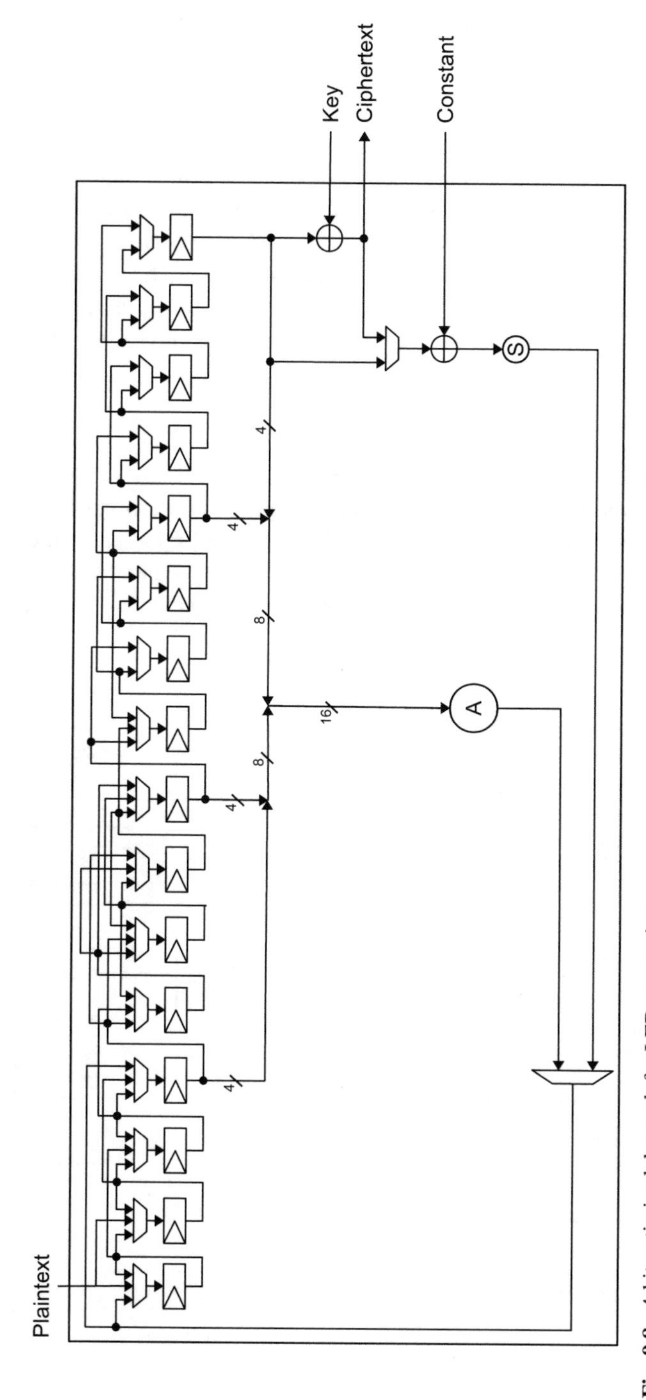

Fig. 9.8 4-bit optimized datapath for LED encryption

of each round, the inverse permutation needs to be implemented. To process a block of data, 30 rounds have to be performed.

Each round is composed of three layers. Only the first layer is nonlinear. It is constructed by a add key step and a substitution step that operates of the first half of the state. The result of this is XORed to the second half and replaces this last half of the state. The sbox is a 4-bit to 4-bit substitution. The second layer, named linear layer, adds some diffusion inside the second half of the state, and finally, the permutation exchanges the two halves of the state and permutes nibbles inside the two halves. This permutation step is not performed during the last round.

The keyschedule is constructed using four linear feedback shift registers (LFSR) that operate 5 nibbles of the key. When all the 80 bits of the key are updated, a subkey is extracted using the current number of round and the same sbox is used for data.

9.4.4.1 Fullwidth Implementation

For the fullwidth implementation of Lilliput, two layers are shared between encryption and decryption because of the Feistel structure. The nonlinear and the linear layers are exactly the same in both modes. It is the case for the permutation because in addition to exchange the halves of the state, an additional permutation operates on the nibble of each half. Figure 9.9 shows the fullwidth datapath.

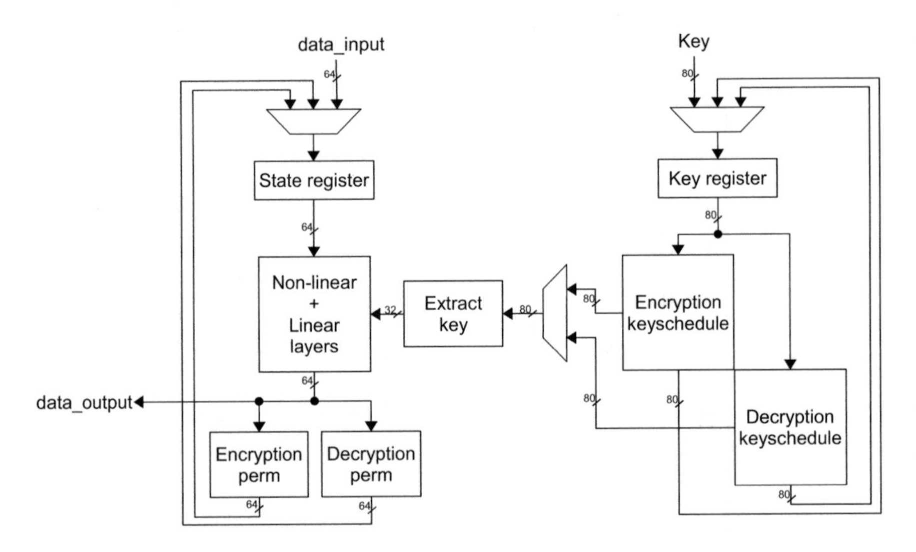

Fig. 9.9 Fullwidth datapath of lilliput

The schedule is non-optimizable and the key part of the datapath follows the general scheme of Fig. 9.3.

9.4.4.2 Serial Implementation

For the serial implementation of Lilliput, the state register is transformed into two shift registers. Additionally, the nonlinear layer and the linear layer are merged to form a single function. The permutation is exactly the same because it requires the full state of data. Figure 9.10 shows the datapath of the serial version of Lilliput.

In Fig. 9.10, all the connections are 4-bit wide except for those whose widths are indicated. As can be seen in Fig. 9.10, the multiplexers on top of the state registers increase in size. Instead of choosing between three elements like in the fullwidth implementation (Fig. 9.9), the serial implementation imposes a first choice between the two permutations and another choice due to the shift. This structure may result in a serial implementation that is not really smaller than the fullwidth implementation of the cipher.

Nevertheless, the nonlinear and the linear layers are really reduced and use only one sbox and 4 4-bit XORs as are 2 multiplexers and 2 additional 4-bit register. The fullwidth implementations use height sboxes and 29 4-bit XORs that operate to a nibble to achieve the same functionality. This may reduce the area of the serial implementation.

The serial hardware implementation uses multiple clock cycles to perform one round of Lilliput. Indeed, based on the construction presented in Fig. 9.10, the serial implementation needs 8 clock cycles instead of 1 to compute one round of Lilliput.

9.4.5 Ktantan

Ktantan is a block cipher with a particular construction that is similar to a stream ciphers. The size of the key is 80 bits and three sizes are available for the size of the block: 32, 48, or 64 bits. This block cipher has been published in 2009 [10].

Because of its particular construction, processing a block of data is time consuming and requires 254 rounds. The state is divided into two shift registers. Two simple combinatorial functions take 6 bits from one of these registers and update the least significant bit of the other register. Additionally, an LFSR with a cycle of 254 is used as round counter and also in one of the two functions. The key is burnt during the operation and 2 bits are generated using the key and the LFSR to serve as round keys.

For the 32-bit version, only one bit per register is updated at each clock cycle. Two bits are updated for the 48-bit version and 3 bits for the 64-bit version. In order to update multiple bits at a time, the basic functions are simply implemented 2 or 3 times.

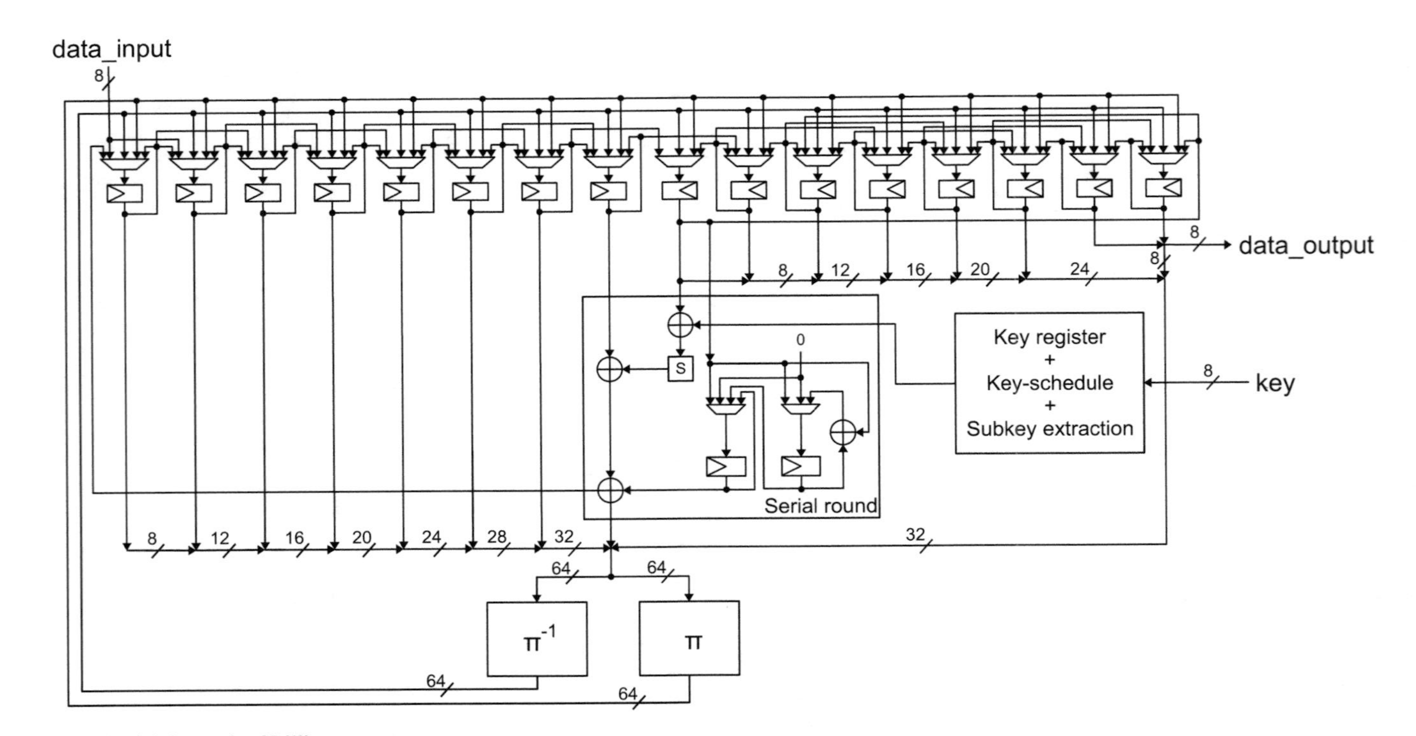

Fig. 9.10 Serial datapath of Lilliput

9.4.5.1 Fullwidth Implementation

Ktantan has a special construction that does not exactly follow the same scheme as the others algorithm. The state of data is split into two parts of different sizes. For the 64-bit version, the two basic functions are applied three times per clock cycle and update three bits at a time. Figure 9.11 shows the encryption datapath of the fullwidth version of Ktantan with the 64-bit state. The keyschedule is only a selection of some bits of the key which is burnt during encryption. This selection uses the round counter as parameter.

Because the two functions (*fa* and *fb*) are very simple, they are non-optimizable and the inverse functions need to be implemented for decryption. The round counter also needs to be reversed in order to provide the correct selection bit for the key.

9.4.5.2 Serial Implementation

Because of the extremely simple structure of this block cipher, the serial implementation is achieved by implementing the function $f(a)$ and $f(b)$ once instead of 3 times. This means the value of the round counter and the value of the key have to be kept

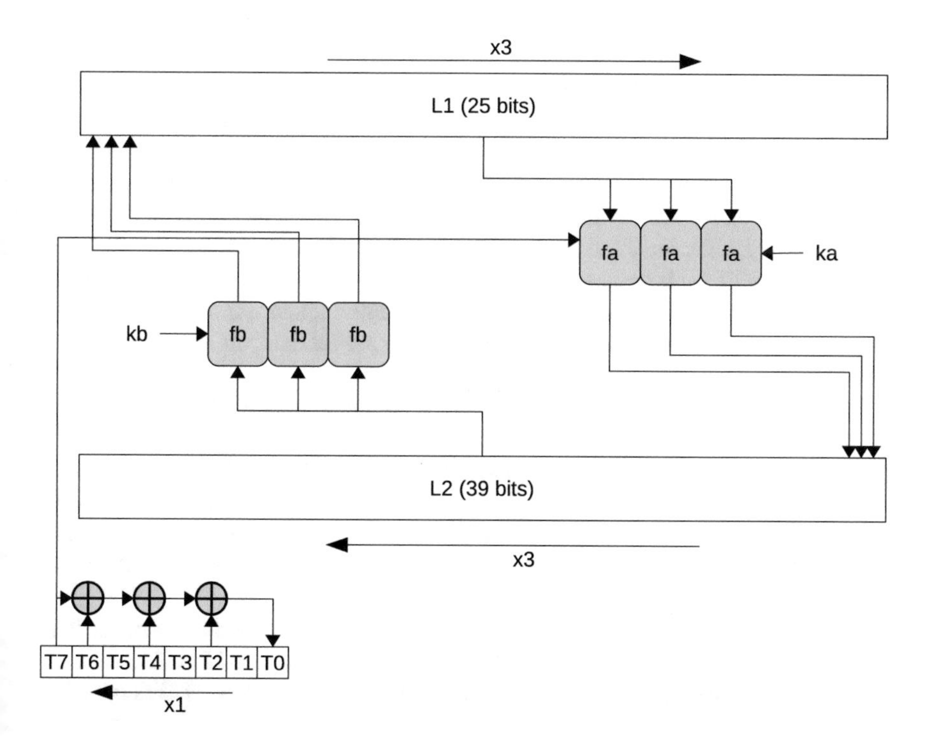

Fig. 9.11 Fullwidth datapath of ktantan

for three clock cycles before to be updated. The state registers are shifted by one bit each clock cycle instead of three.

As a result, a 2-bit binary counter is added, as are some multiplexers. Nevertheless, the area of the serial implementation is expected to be smaller than the fullwidth implementation.

9.5 Comparison of the Implementation Results

To compare all implementations presented in this chapter fairly, a common framework has been developed. This framework uses a simple controller and a 8-bit interface to send and receive data from the cryptographic core. The datapath and the controller are completely separated even if execution of the cryptographic algorithm wrapped in this framework is controlled by the controller. To do so, the only exchange between the controller and the datapath is control signals that ensure the correct functionality of the core that is used.

On the datapath side, the data and the key are separated from the beginning and the output of the block cipher is sent to another 8-bit bus. There is no reset that arrives in the datapath because the entire execution is controlled by the controller and a reset of this last block is sufficient. Additionally, not including a reset in the datapath leads to a smaller design since it is not necessary to add a reset value in the multiplexers on top of the state register.

On the controller side, there is an input of 2 bits that is used to choose the operation to process; this input is named *control* (Fig. 9.12). Three outputs provide information on the current state of the full system. Because of the simple structure of this framework, the data sent at the output of the system correspond to the result of the operation only when the done output of the controller is set to 1.

The choice of 8-bit datapath has been driven by the bigger datapath in all the serial implementation in this work. Reduce this interface to use only 4 bits which may result in better results for bloc ciphers using a 4-bit datapath but this will also lead to unfair comparison.

Figure 9.12 shows the basic schematic of the common framework used to compare the different implementations presented in the previous section.

The areas of the block cipher implementations on FPGAs are usually compared using slices which is the basic block of Xilinx FPGAs. Nevertheless, this metric strongly depends on the family of FPGA used for the implementation. Indeed, two Xilinx FPGAs from different families do not have the same structure and the slices are internally different.

For example, between the Xilinx Spartan 3 and the Xilinx Spartan 6 FPGAs families, look up tables (LUT) that are one of the basic elements of FPGA are totally different as are the entire slices. Xilinx Spartan 3 FPGAs contain two LUTs with four inputs and one output per slice, while Spartan 6 GPGAs contain four LUTs with six inputs and two outputs per slice. Knowing this, it is impossible to compare implementations of block ciphers that have been made on different families of FPGAs

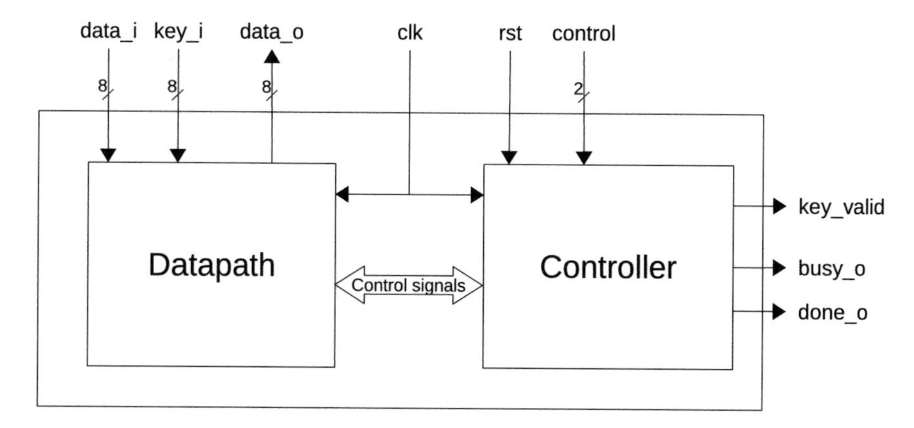

Fig. 9.12 Common framework schematic

using slices as metric. Additionally, a slice is a specific structure of Xilinx FPGAs, which is not applicable with Altera FPGAs.

Nevertheless, almost all SRAM-based FPGAs are constructed using LUTs and flip-flops (FFs). Thus, the best metric to compare two implementations that use the same type of LUTs is to use the amount of FFs and LUTs designs required.

9.5.1 Fullwidth Result

The FPGA that is used in this study is a Xilinx Spartan 6 FPGA (xc6slx16csg324-3). All implementations have been fully tested using post-place and route simulation to ensure that the functionality of all block ciphers implemented is correct for both encryption and decryption modes.

Table 9.2 lists the results of the fullwidth implementations. For each block cipher, the size of the core itself, the size of the controller of the framework, and the entire size of the system are presented.

As shown in Table 9.2, Ktantan is the smallest bloc cipher implemented with only 301 LUTs plus FFs. This is clearly due to its very simple function-based structure. Lilliput is the second smaller bloc cipher with 444 Luts plus FFs. This can be explained by the fact that the encryption and the decryption process share 2 layer over the 3 of this algorithm.

Using the same analysis, of how many parts are shared between encryption and decryption, it is possible to understand why LED is smaller than KLEIN even if it has a key length of 128 bits. Additionally, there is no keyschedule in LED. Nevertheless, those two last ciphers are bigger than Ktantan and Lilliput.

Finally, the number of flip-flops used by each algorithm is very close. Only LED uses a larger amount of flip flop that is explained by the size of the key. Thus, the real comparison here is the comparison of the size of the functions that are employed to

Table 9.2 Results of fullwidth implementations

		LUT	FF	LUT + FF
KLEIN	Controller	24	13	37
	Datapath	604	144	748
	Full implementation	632	157	789
LED	Controller	19	13	32
	Datapath	515	198	713
	Full implementation	549	211	760
Lilliput	Controller	24	13	37
	Datapath	229	144	373
	Full implementation	287	157	444
Ktantan	Controller	17	7	24
	Datapath	143	144	287
	Full implementation	150	151	301

create the cryptographic functionality. This is precious information because it reveals that the round used in LED is more compact than in KLEIN.

9.5.1.1 Serial Implementation Results

For the serial results, the same order is found between the four different block ciphers, the smallest is Ktantan, followed by Lilliput; the two others (LED and KLEIN) are very similar even though KLEIN is slightly smaller. This time, the number of FFs used by each algorithm is completely different. This can be explained by the choices presented for the serial hardware implementations of each block cipher. Each one of them has been carefully designed using specific datapaths.

Table 9.3 presents the results of the implementation of the serial versions of each cipher.

The size of the controller has increased for all ciphers according to the fullwidth implementation results. It is the same for the number of FFs implemented. Concerning the number of LUTs used by these implementations, three of them have seen a reduction: Ktantan, LED, and KLEIN. Only Lilliput increases in the number of LUTs used for the serial implementation. This can be explained by the size of the multiplexers on top of the state register. Indeed, their size has increased because of the permutation function which, in Lilliput, is not optimizable. This function, which uses the full state of data, is thus clearly one of the main drawbacks for any optimization of this algorithm.

Table 9.3 Results of serial implementations

		LUT	FF	LUT + FF
KLEIN	Controller	35	18	53
	Datapath	304	176	480
	Full implementation	330	194	524
LED	Controller	32	19	51
	Datapath	319	198	517
	Full implementation	373	217	590
Lilliput	Controller	35	16	51
	Datapath	235	180	415
	Full implementation	301	196	497
Ktantan	Controller	17	9	26
	Datapath	137	144	281
	Full implementation	137	153	290

9.5.2 *Comparison of the Results*

Table 9.4 compares the results obtained for the complete system for both fullwidth and serial implementation obtained for Xilinx Spartan 6 (xc6slx16).

The algorithm with the larger reduction is KLEIN even if it is still a little bigger than the others. As mentioned just before, the serial implementation of Lilliput leads to a bigger design in Xilinx Spartan 6 FPGAs. The small reduction achieved for Ktantan is explained by the simplicity of the function to be implemented and the small margin left for optimization.

Table 9.4 Results of serial implementations on Xilinx Spartan 6 FPGA

		LUT	FF	LUT + FF
KLEIN	Fullwidth	632	157	789
	Serial	330	194	**524**
LED	Fullwidth	549	211	760
	Serial	373	217	**590**
Lilliput	Fullwidth	287	157	**444**
	Serial	301	196	497
Ktantan	Fullwidth	150	151	301
	Serial	137	153	**290**

Table 9.5 Results of serial implementations using Xilinx Spartan 3 FPGA

		LUT	FF	LUT + FF
KLEIN	Fullwidth	1097	162	1259
	Serial	633	194	**827**
LED	Fullwidth	970	211	1181
	Serial	555	218	**773**
Lilliput	Fullwidth	604	162	**766**
	Serial	638	205	843
Ktantan	Fullwidth	256	151	407
	Serial	222	153	**375**

Results are also generated using Xilinx Spartan 3 (xc3s50) and Table 9.5 shows the area of the two implementations for each algorithm.

In Table 9.5, our results are compared with others studies targeting implementation of block cipher for lightweight application using FPGA. Unfortunately, very few works present results generated using FPGAs for the four algorithms described in this chapter, and only LED has and the comparison is not easy to do. In [3], authors implement the LED block cipher for encryption and decryption but using specific features of the target FPGA. Concerning the results presented in [18], only the encryption is implemented. In these two studies, the LED block ciphers are optimized for area and speed simultaneously. It is nevertheless possible to remark that even if these two other works claims to have very compact implementations, Ktantan is still smaller. Additionally, all the works presented here use only generic components such as LUT and flip-flops for encryption and decryption. That is why the given area for LED is slightly bigger than in [3]. However, the use of BRAM implies that the given area is not the real result and these memory elements need to be taken into account.

9.6 Conclusion and Future Works

This chapter presents the implementation and comparison of four different block ciphers after a quick description of the relatively new field of cryptographic algorithm adapted for lightweight applications. The only constraint explored is the area of the implemented ciphers and a full serial hardware implementation of each of the four algorithms has been presented. The fullwidth hardware implementation, which is the starting point of any hardware implementation work, is also presented. All implementation results are compared fairly, thanks to a common framework. Unlike most of other works in this field, this chapter has presented implementation for both encryption and decryption for all algorithms. If the best candidate for lightweight applications that only needs small area has to be selected from this chapter, Ktantan seems to have the best potential since it is extremely small (290 LUTs plus FFs).

To complete this study, it will be interesting to explore the influence of the interaction between the controller and the datapath on the area of the implementations. Indeed, recent development on this four block ciphers shows that the state encoding may have a strong influence on the implementation results and future researches need to be perform at this point. Additionally, find a way to estimate how much an algorithm can be reduce from the fullwidth implementation which will also be a big improvement in the area-efficient implementation research field. Finally, understand how to deal with the sizes of the different functions such as permutations, keyschedule, and multiplexers is one of the key points to really reduce the area used by any hardware implementation of any algorithm because they usually are the bottleneck for optimized designs as shown in this chapter.

Acknowledgements The authors would like to thank all the CERG team of George Mason university for all the very fruitful discussion and debate during the implementation work.
The work presented in this paper was carried out in the framework of the SALWARE project number ANR-13- JS03-0003 supported by the French Agence Nationale de la Recherche and by the French Fondation de Recherche pour lAronautique et lEspace.

References

1. https://www.cryptolux.org/index.php/lightweight_block_ciphers
2. http://www.univ-st-etienne.fr/salware/block_ciphers_implementation.htm
3. S. Bangari, E. Elavarasi, Fpga implementation of data encryption and decryption using optimized led algorithm
4. R. Beaulieu, D. Shors, J. Smith, S. Treatman-Clark, B. Weeks, L. Wingers, The simon and speck families of lightweight block ciphers. Cryptology ePrint Archive, Report 2013/404, 2013, http://eprint.iacr.org/
5. T. Berger, J. Francq, M. Minier, G. Thomas, Extended generalized feistel networks using matrix representation to propose a new lightweight block cipher: Lilliput. IEEE Trans. Comput. **PP**(99), 1–1 (2015). doi:10.1109/TC.2015.2468218
6. A. Bogdanov, L. Knudsen, G. Leander, C. Paar, A. Poschmann, M. Robshaw, Y. Seurin, C. Vikkelsoe, Present: an ultra-lightweight block cipher, in *Cryptographic Hardware and Embedded Systems - CHES 2007*, ed. by P. Paillier, I. Verbauwhede. Lecture Notes in Computer Science, vol. 4727. (Springer, Berlin, 2007), pp. 450–466. doi:10.1007/978-3-540-74735-2_31
7. L. Bossuet, D. Hely, Salware: salutary hardware to design trusted ic, in *Workshop on Trustworthy Manufacturing and Utilization of Secure Devices, TRUDEVICE 2013* (2013)

8. P. Chodowiec, K. Gaj, Very compact fpga implementation of the aes algorithm, in *Cryptographic Hardware and Embedded Systems - CHES 2003*, ed. by C. Walter, E. Ko, C. Paar. Lecture Notes in Computer Science, vol. 2779. (Springer, Berlin, 2003). doi:10.1007/978-3-540-45238-6_26
9. J. Daemen, V. Rijmen, *The Design of Rijndael: AES - The Advanced Encryption Standard*. Information Security and Cryptography (Springer, 2002). doi:10.1007/978-3-662-04722-4
10. C. De Cannire, O. Dunkelman, M. Kneevi, Katan and ktantan a family of small and efficient hardware-oriented block ciphers, in *Cryptographic Hardware and Embedded Systems - CHES 2009*, ed. by C. Clavier, K. Gaj. Lecture Notes in Computer Science, vol. 5747. (Springer, Berlin, 2009), pp. 272–288. doi:10.1007/978-3-642-04138-9_20
11. T. Eisenbarth, S. Kumar, C. Paar, A. Poschmann, L. Uhsadel, A survey of lightweight-cryptography implementations. IEEE Des. Test Comput. **24**(6), 522–533 (2007). http://doi.ieeecomputersociety.org/10.1109/MDT.2007.178
12. Z. Gong, S. Nikova, Y. Law, Klein: A new family of lightweight block ciphers, in *RFID. Security and Privacy*, ed. by A. Juels, C. Paar, Lecture Notes in Computer Science, vol. 7055 (Springer, Berlin, 2012), pp. 1–18. doi:10.1007/978-3-642-25286-0_1
13. T. Good, M. Benaissa, Aes on fpga from the fastest to the smallest, in *Cryptographic Hardware and Embedded Systems CHES 2005*, ed. by J. Rao, B. Sunar, Lecture Notes in Computer Science, vol. 3659. (Springer, Berlin, 2005), pp. 427–440. doi:10.1007/11545262_31
14. J. Guo, T. Peyrin, A. Poschmann, M. Robshaw, The led block cipher, in *Cryptographic Hardware and Embedded Systems CHES 2011*, ed. by B. Preneel, T. Takagi, Lecture Notes in Computer Science, vol. 6917. (Springer, Berlin, 2011), pp. 326–341. doi:10.1007/978-3-642-23951-9_22
15. D. Hong, J. Sung, S. Hong, J. Lim, S. Lee, B.S. Koo, C. Lee, D. Chang, J. Lee, K. Jeong, H. Kim, J. Kim, S. Chee, Hight: a new block cipher suitable for low-resource device, in *Cryptographic Hardware and Embedded Systems - CHES 2006*, ed. by L. Goubin, M. Matsui, Lecture Notes in Computer Science, vol. 4249. (Springer, Berlin, 2006), pp. 46–59. doi:10.1007/11894063_4
16. F. Karako, H. Demirci, A. Harmanc, Itubee: a software oriented lightweight block cipher, in *Lightweight Cryptography for Security and Privacy*, ed. by G. Avoine, O. Kara, Lecture Notes in Computer Science, vol. 8162. (Springer, Berlin, 2013), pp. 16–27. doi:10.1007/978-3-642-40392-7_2
17. B.J. Mohd, T. Hayajneh, A.V. Vasilakos, A survey on lightweight block ciphers for low-resource devices: comparative study and open issues. J. Netw. Comput. Appl. **58**, 73–93 (2015). doi:10.1016/j.jnca.2015.09.001
18. N. Nalla Anandakumar, T. Peyrin, A. Poschmann, A very compact fpga implementation of Led and photon, in *Progress in Cryptology INDOCRYPT 2014*, Lecture Notes in Computer Science (Springer International Publishing, 2014), pp. 304–321. doi:10.1007/978-3-319-13039-2_18
19. K. Shibutani, T. Isobe, H. Hiwatari, A. Mitsuda, T. Akishita, T. Shirai, Piccolo: an ultra-lightweight blockcipher, in *Cryptographic Hardware and Embedded Systems CHES 2011*, ed. by B. Preneel, T. Takagi, *Lecture Notes in Computer Science*, vol. 6917. (Springer, Berlin, 2011), pp. 342–357. doi:10.1007/978-3-642-23951-9_23
20. T. Shirai, K. Shibutani, T. Akishita, S. Moriai, T. Iwata, The 128-bit blockcipher clefia (extended abstract), in *Fast Software Encryption*, ed. by A. Biryukov, Lecture Notes in Computer Science, vol. 4593. (Springer, Berlin, 2007), pp. 181–195. doi:10.1007/978-3-540-74619-5_12
21. F.X. Standaert, G. Piret, N. Gershenfeld, J.J. Quisquater, Sea: a scalable encryption algorithm for small embedded applications, in *Smart Card Research and Advanced Applications*, ed. by J. Domingo-Ferrer, J. Posegga, D. Schreckling. Lecture Notes in Computer Science, vol. 3928. (Springer, Berlin, 2006), pp. 222–236. doi:10.1007/11733447_16
22. N.I. Technology of Standards, Data encryption standard. Federal Information Processing Standards (FIPS), Publication 46 (1977)

23. T. Suzaki, K. Minematsu, S. Morioka, E. Kobayashi, *TWINE*: a lightweight block cipher for multiple platforms, in *Selected Areas in Cryptography, 19th International Conference, SAC 2012*, Windsor, ON, Canada, 15–16 Aug 2012, Revised Selected Papers, pp. 339–354 (2012). doi:10.1007/978-3-642-35999-6_22
24. W. Wu, L. Zhang, Lblock: a lightweight block cipher, in *Applied Cryptography and Network Security*, ed. by J. Lopez, G. Tsudik. Lecture Notes in Computer Science, vol. 6715. (Springer, Berlin, 2011), pp. 327–344. doi:10.1007/978-3-642-21554-4_19

Chapter 10
Enhancing Secure Elements—Technology and Architecture

Bertrand Cambou

10.1 Introduction

In the past 20 years secure microcontrollers were successfully implemented to create a new class of devices, the smartcards, which have been widely popular to secure handheld terminals in the form factor of SIM cards (Subscriber Identification Module), banking cards, and access cards (ID and transport). About 8 billion of these components are manufactured annually to fulfill the demand. Initially these microcontrollers were manufactured with embedded EEPROM which is now often replaced by Flash to reduce costs. Secure microprocessors, also called "Secure Elements," are in the process of being widely deployed to enhance hardware security of cyber-physical systems (CPS), Internet of Things (IoT), Automotive, smart grid, and many other sensitive systems. Public Key Infrastructure (PKI), an architecture which relies on private cryptographic keys that are kept secret, has been accepted as a way to authenticate users. The success in the adoption of Secure Elements was based on their capability to precisely store these private cryptographic keys for PKI; as well as their ability to perform powerful cryptographic computations such as encryption, decryption, and authentication. The efforts to further enhance secure elements through new Nano-technologies and novel architectures are of strategic importance to reduce cyber-crimes, and to develop a new economy based on trustworthy secure e-commerce.

B. Cambou (✉)
Northern Arizona University, 3452 S. Pimlico Court, Flagstaff, AZ 86005, USA
e-mail: Bertrand.Cambou@nau.edu

L. Bossuet and L. Torres (eds.), *Foundations of Hardware IP Protection*,
DOI 10.1007/978-3-319-50380-6_10

10.2 General Description

Secure elements [1–5] contain integrated microcontrollers with 32–64 bit RISC engine, a crypto-processor with capability to perform encryption/decryption algorithms such as RSA, ECC, AES, and DES. About 10–30 KB of SRAM are usually embedded within the CPU and the crypto-processor. The embedded nonvolatile memories have a capacity of 100 KB to 1 MB and are driven by a secure memory management unit (MMU). The architecture also includes peripheral interfaces, analog and glue logic, power management, and internal clock generators. A simplified block diagram is described in Fig. 10.1. International standards such as ISO/IEC 7816 specify contact-based smart cards, while ISO/IEC 14443 or 18092 specify contactless identification cards. The Global Platform alliance has successfully driven a standardization of the operating systems (OS) along Java-card and Multos. This has allowed the development of third-party value added applications, or Applets, cryptographic methods, deployment of the public key infrastructure (PKI), and client management systems (CMS). Embedded secure elements (eSE) are directly integrated within the final devices, such as a mobile phone. "eSE" does not require the same level of standardization as smart cards, and costs can be reduced by eliminating the micro-module and separate connectors. In wireless applications the business model of eSEs is different than SIM cards, e.g., they are not removable from the phone, so a user cannot change carrier or phone while keeping its SIM card.

10.2.1 *Encryption and PKI Deployment*

Cryptography is an important technology for SEs, we are summarizing some elements as background information for this chapter, [6, 7]. If **P** is the plain text that

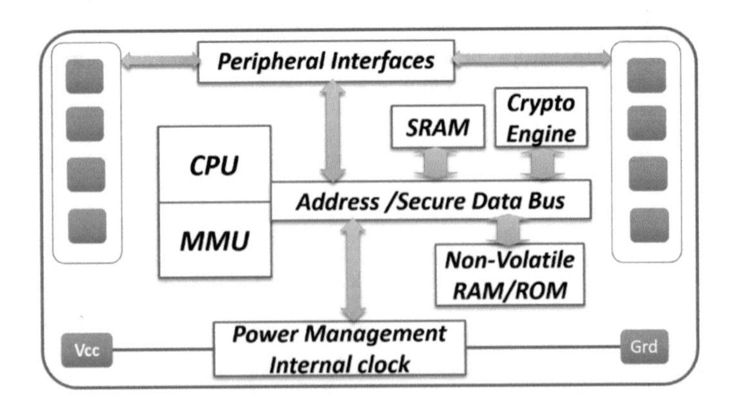

Fig. 10.1 Block diagram of a secure element

need to be encrypted, $\mathbf{K_S}$ is the symmetrical key, and **E** is the encryption algorithm, the cypher **C** is given by the Eq. (10.1), and can be decrypted with algorithm **D** and the same key:

$$\mathbf{C} = \mathbf{E}\left(\mathbf{P},\ \mathbf{K_S}\right) \tag{10.1}$$

$$\mathbf{P} = \mathbf{D}\left(\mathbf{C},\ \mathbf{K_S}\right) \tag{10.2}$$

The Data Encryption Standard (DES) algorithm based on 56 bit keys, developed by IBM and the US government in the 70's, is inadequate for high security applications. It has been replaced frequently after 2000 by the Advanced Encryption Standard (AES) algorithm that is based on 128 bit keys or larger, as developed by Joan Daemen and Vincent Rijmen; AES is also called the Rijndael code. Both DES and AES algorithms are usually ported into secure elements to perform basic encryption/decryption function. Symmetrical encryption methods are very effective and play an important role in hardware authentication. However, these methods are not appropriate to handle large groups of users. The invention of asymmetrical cryptography in the 70's has been widely adopted for this purpose to secure CPS, mobile and fixed users, access control, and IoT. Each user (**i**) has two unique cryptographic keys, a public key $\mathbf{K_{PU\text{-}i}}$ that is shared with other users, and a private key $\mathbf{K_{PR\text{-}i}}$ that is kept secret (*Note: in the case of IoT, and secure elements, both private and public keys are kept secret*). These two keys have a level of symmetry with each other: It is possible to encrypt with one key, public or private, and to decrypt the result with the other key. Two cyphers **C1** and **C2** can be created from these two keys following the equations:

$$\mathbf{C1} = \mathbf{E}\left(\mathbf{P},\ \mathbf{K_{PU-i}}\right);\quad \mathbf{P} = \mathbf{D}\left(\mathbf{C1},\ \mathbf{K_{PR-i}}\right) \tag{10.3}$$

$$\mathbf{C2} = \mathbf{E}\left(\mathbf{P},\ \mathbf{K_{PR-i}}\right);\quad \mathbf{P} = \mathbf{D}\left(\mathbf{C2},\ \mathbf{K_{PU-i}}\right) \tag{10.4}$$

The method allows two users (**i**) and (**j**) to communicate with each other after exchanging their public keys; the user (**i**) can send the cypher **Ci** to user (**j**) who can decrypt it to retrieve **P**:

$$\mathbf{Ci} = \mathbf{E}\left[\mathbf{E}\left(\mathbf{P},\ \mathbf{K_{PR-i}}\right),\ \mathbf{K_{PU-j}}\right] \tag{10.5}$$

$$\mathbf{P} = \mathbf{D}\left[\mathbf{D}\left(\mathbf{Ci},\ \mathbf{K_{PR-j}}\right),\ \mathbf{K_{PU-i}}\right] \tag{10.6}$$

In this double encryption method, users (**i**) and (**j**) keep their private key secret from each other. This can offer multiple benefits to the communication:

- *Electronic (digital) signature of user (**i**).* User (***j***) knows that the message originates from (***i***) because it is assumed that only (***i***) possess the private key $\mathbf{K_{PR\text{-}i}}$. So, **E (P,** $\mathbf{K_{PR\text{-}i}}$**)** can be considered as an electronic signature.
- *Protection of the user (**j**).* The only way to decrypt **Ci** is to possess the private key $\mathbf{K_{PR\text{-}j}}$ that is kept secret by user (**j**).

- *Non-repudiation.* To seal the transaction user (**j**) has to send back to user (**i**) an encrypted digital signature.

The commercial deployment of asymmetrical cryptography is called public key infrastructure, (PKI). Secure elements were instrumental in the deployment of PKI to enhance access control. In the late 70's Ronald Rivest, Adi Shamir and Leonard Adleman developed a beautiful algorithm carrying their name, RSA [7], that provides an effective asymmetrical encryption and decryption method. The method exploits the number theory, and in particular the Euler–Fermat theorem; the private and public keys are inverse modulo numbers. To this date the algorithm has been proven as unbreakable; powerful quantum computers might challenge this statement in the future. RSA tends to be highly computing intensive, and several orders of magnitude slower than AES. This is no longer a limitation as the compute power of secure elements is now appropriate. Alternate asymmetrical algorithm such as those based on elliptic curves (ECC) are much lighter than RSA, and are widely deployed on the cost-sensitive versions of secure elements. In order to strengthen both symmetrical and asymmetrical cryptography, it is important to add functions **H** based on random numbers generators (RNG) to the plain texts **P**. For example Eq. (10.1) and (10.2) can be written the following way:

$$\mathbf{C} = \mathbf{E}\,(\mathbf{P} + \mathbf{H},\ \mathbf{K_S}) \tag{10.7}$$

$$\mathbf{P} + \mathbf{H} = \mathbf{D}\,(\mathbf{C},\ \mathbf{K_S}) \tag{10.8}$$

In this example both **P** and **H** are described by a binary stream of data, and **P** length is known by the receiving party. **P + H** is much longer and complicated to decrypt than **P** to prevent third-party attack. Secure elements need to have a high quality RNG integrated into the component to encrypt important information.

What to remember: *Secure elements are an essential part of the effective deployment of PKI. They act like vaults hiding cryptographic keys, the private keys for RSA or ECC, and the symmetrical AES keys. Secure embedded non-volatile memories (NVMs) that store these keys are therefore critical. The crypto-processors that decrypt and encrypt messages need access to strong RNGs.*

10.2.2 Multi-function Authentication

In asset, and people protection there is a major difference between "*identification*" and "authentication. "*Identification*" does not have to be secret as long as they are clear, unique, and unambiguous. "Authentication" is a way to prove that the person or the object is the right one. Authentication factors have to be absolutely secret, however, they do not have to be unique as long as they can offer a high level of certainty of a match between the person or object involved in the transaction and the ones expected. For example a nice picture with a first and last name are good

Table 10.1 False rejection rates (FRR); False acceptance rates (FAR)

Response to test	Correct ID	False ID
Positive authentication	***Correct***	*False (high FAR)*
Negative authentication	*False (high FRR)*	***Correct***

identification factors, but are not secret and cannot be authentication factors; pin codes are not unique, and cannot be an identification factor, however can be an acceptable authentication factor when kept confidential. In most e-transactions the subject has to provide its ID number as well as the secret password for authentication. The authentication factors are the first line of defense to prevent hackers from breaking a system. Both identification and authentication factors are stored permanently in secure elements, the users have to provide fresh authentication factors when they wish to access a service. The full authentication cycles are done within the secure elements to enhance secrecy [8, 9].

Authentication factors include pin codes, passwords, biometric prints such as finger print, iris, blood vessels in the fingers, facial features, and voice, as well as hardware prints such as PUFs that exploit micro variations occurring during integrated circuit fabrication. The use of multiple factors, also referred to as multi-function (or multi-factor) authentication is expected to be stronger than a single authentication that is based on only one factor. Trustworthy authentications incorporate cryptography to increase the secrecy of system.

Authentication methods have to reduce both the false rejection rates (FRR) resulting in false negative authentication, and the false acceptance rates (FAR) resulting in false positive authentications, see Table 10.1. False negatives are frequent with biometry considering that human-based prints are not constant. Fingerprint authentication frequently faces high FRR. Error correction is shown to reduce FRR [10, 11]. However, may also correct hostile messages thereby increasing FAR.

What to remember: *SEs are utilized to provide reliable multi-function authentication for access control. The metrics to quantify the quality of the authentications are FRR for false negatives, and FAR for false positives.*

10.2.3 Embedded Secure Memories—Utilization of Flash

Embedded memories of secure elements (SE) have to store client-related databases, all cryptographic and authentication keys, the operating systems, and act as cache memory for the processor. The size of the embedded memories varies from a few k-bits for ID and RFID applications to the M-byte level for high-end applications. At low densities EEPROM is the preferred technology due to its excellent array efficiency. For low power and low density applications ferroelectric memories, FeRAMs, secured a niche market, and have the potential to expand. The dominant technology for SEs remains embedded flash [12, 13], this is due to its low cost at

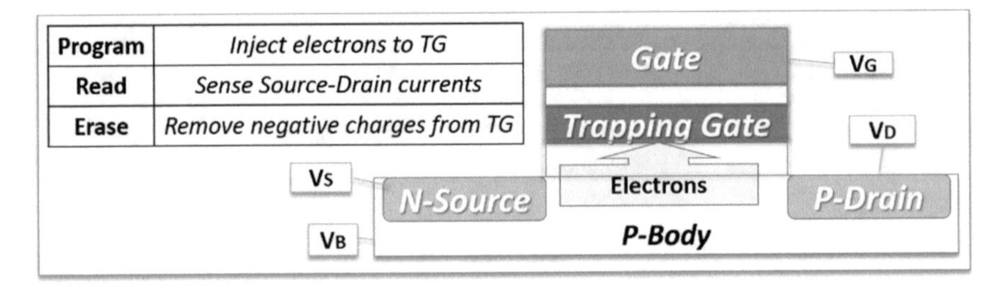

Fig. 10.2 Flash cell

high density, and ability to shrink to the 32 nm lithographic node. NOR has been the architecture of choice for embedded flash, however embedded NAND was also successfully introduced to reduce the size of the arrays. A cross-section of a flash cell is presented Fig. 10.2.

The basic principle of flash is the modulation of the threshold voltage of the metal–oxide–semiconductor (MOS) transistor by selectively trapping electrons in a floating/trapping gate (TG). "1s" are programmed by injecting electrons in the TG, thereby increasing the threshold voltage of the transistor. In NOR flash the electrons are injected through the thin dielectric, also called tunnel, using channel hot electrons (CHE). The energy to "heat" the electrons is added by applying at least 3.5 V between source and drain, the injection occurs near the drain. In NAND flash the electrons are directly injected by Fowler Nordheim (FN) effect through the tunnel oxide. If **ΔV** is the voltage differential across the tunnel dielectric, **Tox** its thickness, **A** and **B** are physical parameters, the current **J** circulating through the dielectric is

$$\mathbf{J} = \mathbf{A}\left(\frac{\mathbf{\Delta} V}{Tox}\right)^2 e^{-B/\left(\frac{\Delta V}{Tox}\right)} \tag{10.9}$$

The tunnel dielectric has to be thick enough, above 5 nm, to retain information and prevent charge losses, so the voltage V_{BG} necessary to program a NAND flash cell has to be large enough, usually greater than 15 V. Both CHE and FN require high voltage transistors that are expensive to integrate with high density logic. The programming cycles are done step by step to avoid over-programming conditions. Multi-bits per cell are commonly obtained by modulating the electrostatic charges in the TG, thereby creating multiple levels of threshold voltage. For 2-bits per cell, 4 levels are needed, for 4-bits per cell, 16 levels are needed. Traditional Flash cells use floating polysilicon gate to trap these electrons. There is a growing interest to replace the polysilicon with silicon rich silicon nitride TG to reduce the thickness of the structure, and cell to cell electrostatic interactions for high density flash memories. Mirror bit technology can get 2-bits per cell by trapping charges in a non-conductive nitride TG, one bit near the source, the other one near the drain.

Table 10.2 Vulnerability analysis

Analysis	Description	Flash vulnerability
DPA/SPA	Differential Power Analysis "0" and "1" drive different currents	• Highly vulnerable during read cycles • Also vulnerable during programming
EMI	Measure the currents on the bus connecting the memory blocks	• Very effective on Flash. Magnetic shielding is rarely used
Physical attacks Delayering, SEM	Electron microscopy can extract the charges trapped in the cells	• Charges in the TG deflect e-beams • Electrons trapped in the thin oxides
Thermal decoration	Cold spray (liquid nitrogen) decorate hot spots in the memory	• Differentiate "0" and "1" after Read cycle due to the high currents

The read cycles are based on sensing the source drain currents that are different for "0"s and "1"s. During erase cycles the negative charges are removed from the trapping gate by FN effect to return to the original threshold voltage. Typically embedded flash memories have an endurance of 100,000 program-erase cycles.

Vulnerability and limitations of flash memories: Crypto-analysts and hackers developed very effective methods and side channel analysis' to extract the content of flash memories. In most cases they are exploiting the fundamental physical properties of the cells and the way they operate. The protection of the information stored in secure elements is critical for the integrity of cryptographic operations. A summary of the vulnerability of flash is shown on Table 10.2.

- *Differential Power Analysis (DPA)*: DPA or single power analysis (SPA) is based on the measurement of the current on ground PIN during operation with a fast signal analyzer [14]. During the read cycles the "0"s and "1"s drive different currents that are visible with DPA, thereby exposing confidential information, and keys.
- *Electromagnetic interferences analysis (EMI)*: The current circulating through the data bus which connects the memory and the processor is measured by a magnetic sensor placed above the chip, and a signal analyzer. Like DPA, the method differentiates the "0"s from the "1"s.
- *Physical attacks*: The passivation and the metal layers are removed to allow Secondary Electron Microscopy (SEM) to detect the charges trapped in the flash cells differentiating the "0"s and the "1"s. The method also allows the detection of the number of program-erase cycles which leave behind charges trapped within the tunnel dielectric.
- *Thermal decoration*: Liquid nitrogen allows the deposition of ice on the chip. The hot spots where electric current circulate melt and decorate the ice. The high power flash technology is vulnerable to these attacks.

Flash memory has also the following limitations for secure elements:

- *High voltage, high power*: The programming-erase cycles of flash devices need high operating voltages (10–18 V). This is challenging for applications, such as RFID, that are power sensitive.
- *Slow operation*: Flash is about 5X slower to read than other memories, and orders of magnitude slower to program. This could open opportunities for the crypto-analyst due to the slowness the crypto-processor.
- *Complex manufacturing*: About 10 additional masking levels are needed to manufacture embedded flash products above and beyond of the basic CMOS flow. It is doubtful that embedded flash will be the preferred solution for secure processors below the 28 nm node. Candidate technologies to replace flash in the embedded space include ReRAM, and MRAM. The usage these technologies for SEs is covered in the following sections.

What to remember: *Embedded flash is currently the mainstream technology for mid- to high-end secure elements. However, crypto-analysts are now armed with effective methods to extract the confidential information stored in flash memories.*

10.3 Usage of Advanced Memory Technologies

10.3.1 Comparison with Resistive RAM, and Magnetic RAM

ReRAM cells [15–17] are based on Nano-materials inserted between two electrodes switching back and forward between Low Resistivity States (LRS) for “0s,” and High Resistivity States (HRS) for “1s.” Nano-materials such as metal oxides behave as solid electrolytes. As it is shown in Fig. 10.3 the basic principle behind ReRAM, and the switch between LRS and HRS, is the reversible formation of highly conductive filaments between the electrodes. These conductive filaments consist of positively charged oxygen vacancies, or metal ions, migrating toward the cathode. The power necessary to operate ReRAM cells is small compared with flash, and the switching times are lower. Current industrial R&D investments in ReRAM are massive, and this technology is expected to play a major role for future SEs. Unlike flash, ReRAM can be embedded at advanced CMOS nodes without adding too much complexity, due to its compatibility with low power and mainstream manufacturing processes.

Magnetic RAM: Like ReRAM cells, [18], MRAM cells can switch back and forward between the two states HRS and LRS, see Fig. 10.4. The cells consist of two magnetic domains, a fixed reference, and a second domain that can be programmed in two different directions. A tunnel layer usually made of thin Magnesium or Aluminum Oxides separates the two domains. When the magnetization of the two domains is aligned, the resistivity of the tunnel oxide is low (LRS) due to the giant magnetoresistance effect; when anti-aligned the resistivity is high (HRS).

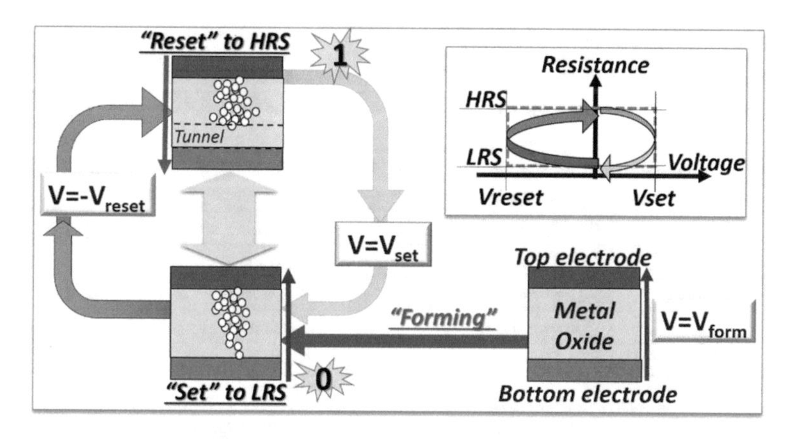

Fig. 10.3 ReRAM cell

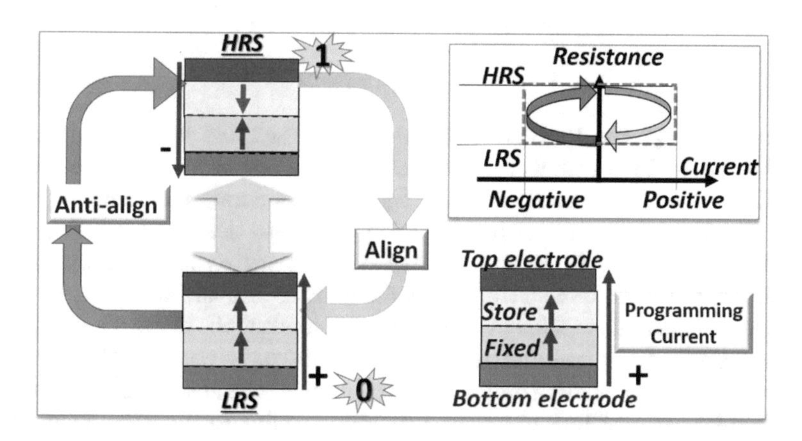

Fig. 10.4 Vertical STT MRAM cell

In the example shown in Fig. 10.4, the domains have vertical magnetization, and the programming method called STT or spin torque transfer is based on the circulation of a current through the structure. Other MRAMs are based on horizontal magnetization and the circulation of a separate current to program the cells through magnetic field. The small ratio HRS/LRS for MRAMs (about 2 for MRAM compared with 100 or more for ReRAM) is hard to master in volume manufacturing.

Back-to-back comparison: The typical parameters for embedded flash, ReRAM, and MRAM are compared in Table 10.3. For each technology there are large variations, depending on the lithography nodes, the suppliers, and the type of structures. Flash data is the most reliable because ReRAM and MRAM are not yet running in volume production. However, it is clear that embedded flash memories are slower and operate at higher power.

Table 10.3 Typical parameters flash—ReRAM—MRAM

Operation		Flash	ReRAM	MRAM
Program	Physics	Inject charges	Form filament	Domain orientation
	Parameter	NOR $V_{DS} = 5$ V; NAND $V_{GB} = 15$ V	Vset = +2 V	Current: 500 μA
	Power	1 mj/bit	10 pj/bit	100 pj/bit
	Speed (ns)	5,000 ns/block	2–20 *ns*	2–20 ns
Read	Physics	S-D current	Resistance	Resistance
	Parameter	Voltage: 10 mV	Current: 1–20 μA	Current: 1–20 μA
	Power (pj)	10 pJ	1 pJ	1 pJ
	Speed (ns)	50 ns	2–20 ns	2–20 ns
Erase	Physics	Remove charges	Program LRS	Program LRS
	Parameter	$V_{GB} = 15$ V	Vreset = −2 V	Current: 500 μA
	Power (pj)	10 μJ/bit	10 pJ/bit	100 pJ/bit
	Speed (ns)	1,000 ns/block	2–20 ns	2–20 ns

- *Programming*: The ReRAM cells can switch back and forward between the HRS and the LRS with a few volts, μA currents, and below 10 ns. In MRAMs both STT, and TAS need relatively high currents during programming cycles. Flash can run into an over-programming cycle, and needs to be erased before reprogramming. ReRAM and MRAM do not need erase cycles before programming. For all three technologies the number of guaranteed program-erase cycles is greater than 100 thousand cycles, with retentions in excess of 10 years.
- *Read*: Unlike flash, both ReRAMs, and MRAMs can be read quickly at constant current, thereby minimizing exposure to side channel attacks. The high HRS/LRS ratio of ReRAM is particularly attractive to minimize read errors. With flash, charges get trapped in the tunnel oxide during program-erase cycles, so the threshold voltages are drifting faster with the cells that are frequently switched from the two states; the cells storing cryptographic keys tend to be re-programmed less often, and could stick out during crypto-analysis. At its forefront, it has been claimed that both ReRAM and MRAM can reach read access times of 1 ns versus 50 ns for flash.
- *Erase*: It is possible to quickly erase ReRAM at low voltage, low current, and in a few ns. This feature is important when a real-time attack of the SE is detected. With flash, crypto-analysts can clamp down Vcc below 1.5 V during side channel attacks preventing a defensive erase cycle and freely analyze the device. During the life cycle of SEs it is also required to reprogram some of the cryptographic keys, this is faster with ReRAMS.

What to remember: *The back to back comparison of flash, ReRAM, and MRAM for SEs is favorable to ReRAM due to its low power and fast access time. Flash benefit from its legacy position for now, and MRAM lacks manufacturability.*

10.3.2 Usage of Content Addressable Memories (CAM)

Internet Protocol (IP) routers and high-performance microprocessors are currently using CAM to accelerate the rate of pattern matching per second. Large CAMs contain data bases of IP addresses that are compared at once with one particular IP address, the input pattern. CAM has a parallel architecture that extracts the location in the memory of these matching IP addresses. The comparison of the input pattern with thousands of IP addresses are done "in situ" without ever reading the input pattern. At the level of each cell, the stored bits are XORed (XOR is the Boolean instruction $\oplus$ defined by: $\mathbf{0 \oplus 0 = 0}$; $\mathbf{1 \oplus 1 = 0}$; $\mathbf{0 \oplus 1 = 1}$; $\mathbf{1 \oplus 0 = 1}$) with the bits from the input pattern. Each CAM cell typically contains two 6T SRAM cells, one for the stored bits, one for the input bit, and an XOR logic element.

Beside SRAM, there are methods to design a CAM architecture with flash, ReRAM or MRAM [19–21]. Figure 10.5 shows an example of CAM cell design with ReRAM. In this example two pin codes are stored in the ReRAM. During cycle-1 of the authentication, the four XOR engines (C1, C2, C3, C4) compare pin-1 (the stored pattern is U**1110**U) with the input data U**0010**U; then pin-2 U**0010**U is compared during cycle 2 of the authentication with the same input data, this is the matching pin. In case of a mismatch the parallel architecture will not disclose which bit is a miss. This design, of Fig. 10.5, can be generalized by increasing the width of the compare elements to words as large as thousand bits

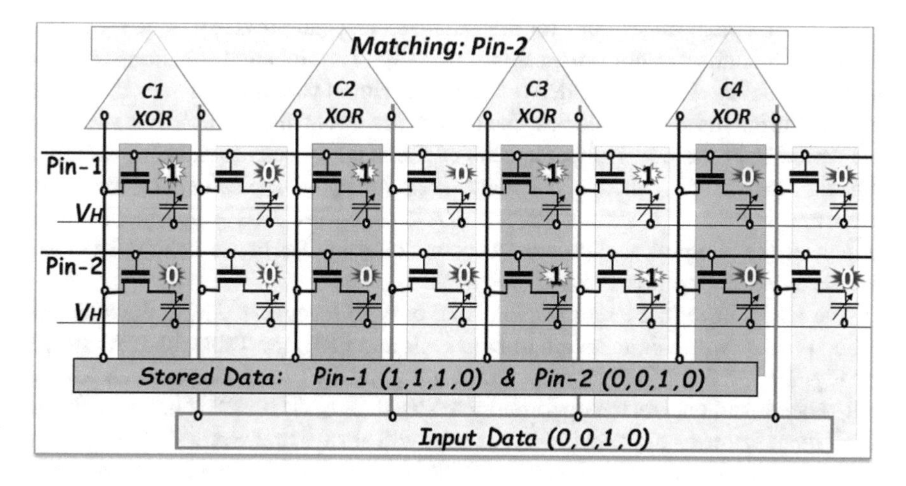

Fig. 10.5 CAM architecture with ReRAM

rather than 4, and by adding parallel elements to perform multiple compare cycles at once. When embedded within secure elements, the CAM architecture can be very effective in performing some important authentication functions such as password check, pin code verification, or PUF challenge-response pairs (CRP) matching. This method can resolve a weakness inherent with traditional RAM architecture where password or pin codes are extracted from the memory array for authentication, exposing the information to crypto-analysts. One of the side channel analyses to extract secret keys is to bring bad keys to the secure element, and to observe the transfer of data between the memory block and the processing element. The ability to directly compare an input key with stored patterns in a CAM eliminates this potential attack. This solution is not applicable for all cryptographic primitives. For example private key needs to be extracted from the secure memory to decrypt messages encrypted with the public key. It is then desirable to have both a RAM, and a CAM architecture available in the SE to manage the multiple keys and cryptographic primitives:

- *Storage in the embedded RAM*: asymmetric keys (RSA, ECC public and private keys), symmetric keys (DES, AES…), biometric minutia.
- *Storage in the CAM*: pin codes, passwords, PUF responses, biometric print.

What to remember: *Authentications such as* in situ *password verification can benefit from CAM architectures and reduce the exposure to crypto-analysis. RAM architecture is still needed for cryptographic key storage.*

10.4 Usage of Physically Unclonable Functions

10.4.1 PUFs Within Secure Elements

PUFs act as virtual fingerprints for the hardware, and can provide unique signatures during the authentication processes [22, 33]. The inherent randomness, and uniqueness of the PUF can be derived from the natural variation obtained during the manufacturing of memory blocks. The underlying mechanism of a PUF is to exploit the variation of the physical parameters of the memory cells to generate an initial digital "fingerprint" called a challenge (i.e., input) of typically 128 bits. This challenge is sent to a secure server for reference. The PUF can generate fresh responses (i.e., output), and produce a set of Challenge-Response-Pairs (CRPs) that are matching during positive authentications.

The CRPs have to be reproducible, and easy to recognize during the authentication process with minimization of both FAR and FRR, see Table 10.1. As shown on Fig. 10.6, 6T SRAM memory arrays are good PUF candidates. Each SRAM cell, when tuned on, has the opportunity to switch as a "0" or a "1," however due to small manufacturing asymmetries some cells will always prefer one side. As a result a PUF array consisting of multiple SRAM cells can generate reproducible

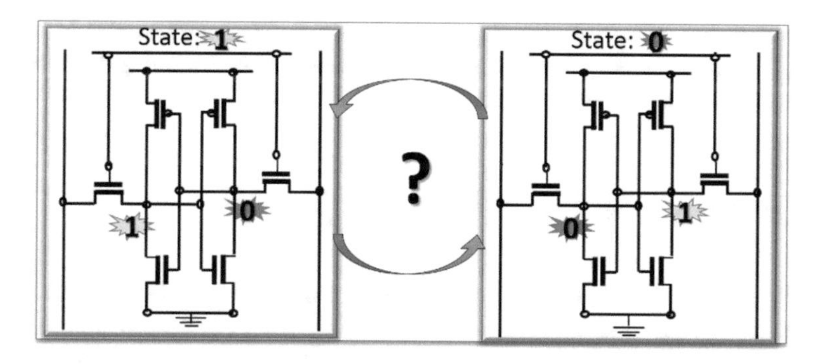

Fig. 10.6 PUF with SRAM

challenges and responses. This method has been exploited for commercial authentications. Such PUFs have a number of limitations and lack trustworthiness that could create a false sense of security. One major concern is the exposure to hacker, and their ability to extract un-encrypted PUF responses, which largely defeats the value of the method. There is value to embed the PUF within a secure encrypted elements that can decrypt the challenges coming from the secure server, see Sect. 10.2.1, and thereby perform "in situ" authentication. Below is an example of authentication with secure element:

- *Assumption 1*: the SE **A** generates a PUF challenge C_A that was downloaded to the server **B**, it has RSA keys ***KprA*** and ***KpuA***, an identification ID_A, and an AES key ***Ks***. The PUF generates responses to ***RAi*** for each authentication "*i*," and hash functions HAi_1 and HAi_2 based on its own RNG.
- *Assumption 2*: the secure server **B** has the keys ***KprB***, ***KpuB***, and ***Ks***, as well as a data base with the information related to **A** that includes C_A, ***KpuA***, and ID_A, and can generate the hash function ***HBi***.

In this example the authentication is done in three steps: Step 1 the SE **A** sends its encrypted identity ID_A *to* the server **B**; Step 2 **B** sends an encrypted challenge C_A to **A**; Step 3 **A** performs the CRP matching to authenticate **B**, if positive **A** send back to **B** the certificate $ID_A + C_A$:

$$\mathbf{C1 = E((E\{ID_A,\ Ks\}) + HAi_1,\ KpuB)} \tag{10.10}$$

$$\mathbf{C2 = E\,[E\,(E\{C_A,\ Ks\} + HBi),\ KprB],\ KpuA]} \tag{10.11}$$

$$\mathbf{C3 = E\,[E\,(E\{\{ID_A + C_A,\ Ks\} + HAi_2,\ KprA),\ KpuB]} \tag{10.12}$$

Such an authentication protects for both parties, **A** identifies **B** with the PUF, and **B** identifies **A** with its ID and asymmetrical key. The concept of non-repudiation was incorporated in this communication, because both the SE and the server did use their secret private key in the process.

What to remember: *Embedding PUFs, which are acting as hardware "fingerprints," within secure elements strengthens the authentication processes, and allows all communication between the SE and the server to be encrypted.*

10.4.2 PUFs with Embedded Memories

In addition to SRAM, PUFs can be generated from DRAM, ReRAM, MRAM, and flash memories, see Table 10.4. Each memory technology has physical parameters that can be exploited for the design of PUFs. Variations of properties such as lithographic critical dimensions, doping levels of semiconducting layers, resistivity of connecting materials, threshold voltages of MOS transistors, and others [23–33] can make each PUF device unique and identifiable from all others when produced by the same manufacturing process.

- *DRAM*: In a DRAM, each cell contains a capacitor that is selectively charged to store a "1" (yes) or a "0" (not charged). Subject to natural leakages, these cells need to be recharged during the refresh cycles, typically every 1 ms. One way to create a PUF is to program an array, then measure the voltage left after a fixed amount of time, for example 2 ms. For challenge and response generation, the bottom 50 % of the distribution of the cells that leaks the most can arbitrarily be "0"s, while the other half "1"s. This method can be used for standalone DRAM, not with existing SEs.
- *ReRAM*: the creation of PUFs out of ReRAM memories is still in a research mode. A possible way is to use built-in-self-test (BIST), ref [34] to test a particular parameter in the ReRAM cells of the array, and determine the "0" from the "1". Three parameters are candidates for PUF, **Rmax** corresponding to HRS, **Rmin** corresponding to LRS, and **Vset** (see Fig. 10.3). **Rmax** exhibits the largest cell to cell variation, while **Vset** has a tighter distribution. **Rmax**-based PUFs are expected to have the lowest CRP error rates, FAR, and FRR. **Vset**-based PUFs might have higher error rates, however, they are more difficult to

Table 10.4 PUF generation with embedded RAM

Memory	Example of PUF generation	Quality
SRAM	*Random Flip of the 6T cell: start as a "0" or a "1" after power up*	*Mainstream but not really secure*
DRAM	*Discharge the capacitors, then measure: Get a "0" or a "1"*	*Need constant refresh*
ReRAM	*Variations of the value of the Vset: Define a "0" or a "1"*	*Quite novel*
MRAM	*Variations of the Rmax's after programming: Define a "0" or a "1"*	*Quite novel*
Flash	*Partial proprammina, then measure: Get a "0" or a "1"*	*Slow programming*

extract through side channel analysis because the "0"s and the "1"s can be close to each other.

- MRAM: The method to create PUFs with MRAM is similar to that of ReRAMs. The resistivity of the HRS can be exploited, with a transition between the "0"s and the "1"s, located at the median of the distribution.
- *Flash*: a method to create PUF from flash memories is to exploit the programming mechanism. As shown in Fig. 10.2, during flash cell programming electrons are injected to the trapping gate to change the threshold voltage of the MOS transistor. In order to generate PUF challenges and responses, the injection time is fixed, giving each cell the opportunity to have different threshold voltage; 50 % of the cells with the lowest threshold are set as a "0," the remaining are "1s."

Regardless of the type of memory technology, the unique signatures of the PUFs are derived from their intrinsic manufacturing variations, which occur during the fabrication process. There is an expected level of mismatches between the distribution of parameter P when the challenges were generated (i.e., the initial "print" that was stored in the secure server), and when the responses are freshly generated (the "print" generated before authentication), see Fig. 10.7, where parameter **P** is the **Vset** measured on ReRAM samples manufactured at Virginia Tech. The solid electrolytes of these samples are made of CuOx. The median of the distribution is $\mathbf{\mu = 2.1\ V}$ with a standard variation $\mathbf{\sigma = +0.54\ V}$. When a cell has a **Vset** below $\mathbf{\mu}$, a"0" is generated as a challenge, "1"s are generated when **Vset** is above $\mathbf{\mu}$. The second graph called "next PUF" in Fig. 10.7 is an example of distribution that has drifted between challenge and response.

The drifts occurs when the PUFs are subject to changes related to temperature, voltage, EMI, aging, and other environmental factors; the resulting responses are then different than the original challenges. The potentially undesirable consequences are weak PUFs with high CRP error rates, and high false rejection rates

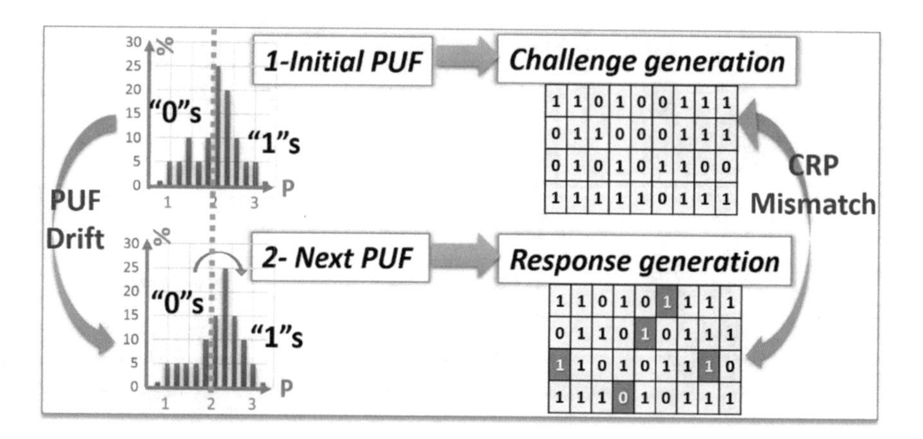

Fig. 10.7 Challenge response pair errors

(FRR). Error correction could reduce CRP errors, however, they could also blindly correct hostile challenges, and increase false acceptances (FAR).

What to remember: *PUFs can be generated by most commercial memories, namely SRAM, DRAM, ReRAM, MRAM, Flash, FeRAM, and EEPROM. During authentications, the PUF responses can drift away from the initial challenges of reference, creating CRP errors as well as generating FRRs and FARs.*

10.4.3 Strengthening Secure Memory-Based PUFs with Ternary States

PUFs, in addition to having low FRR and low FAR, should be **nonobvious** and **unclonable**, to prevent a third party from easily extracting the responses. These objectives are somewhat opposite, easy to extract parameters P that could fulfill the low CRP error rate but fail the obligation of nonobviousness, and this if they are easy to extract by hackers. Conversely, PUF could be extremely hard to uncover by crypto-analysis at the cost of being hard to read for challenge/response generation, thereby having high CRP error rates and unacceptable FRRs. In the following two sections, we present several methods that are aimed to achieve concurrently these objectives, and analyze their potential based on experimental results.

One suggested method that uses ternary states, "0," "1," and "X" is shown above, Fig. 10.8. In this method the cells with parameter P close to the transition are blanked "X," [35]. During challenge generation, only the streams of "0," and "1"s that are solid are sent to the secure server as PUF challenges. The positions of the cells that are "X" are kept as reference in the memory. During response generation, only the cells that are not blanked as "X" are tested to generate the binary stream

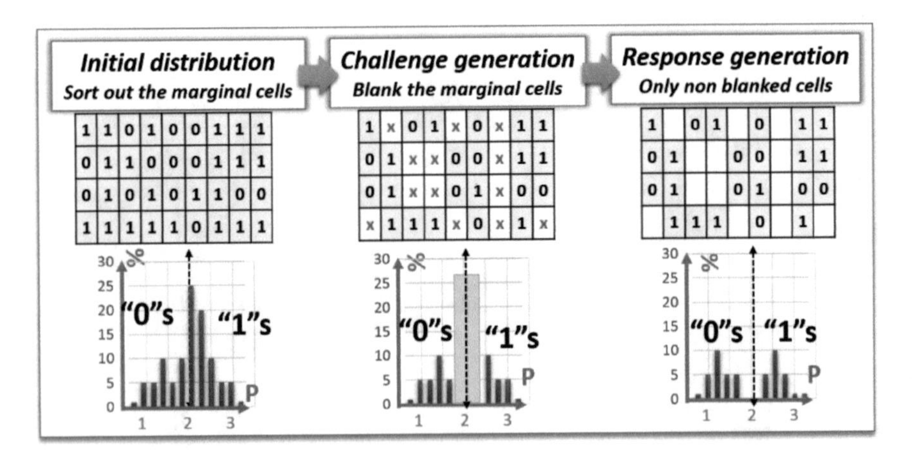

Fig. 10.8 Challenge-response generation with ternary states

used for authentication. The likelihood that a cell previously tested as "0" or "1" during the challenge generation can flip is low.

To study the robustness of the method, we conducted a statistical experiment using the same ReRAM samples presented in Figs. 10.7 and 10.8. We characterized the **Vset** distribution for several individual ReRAM cells. These particular cells were subjected to repeated reset and set operations under the same conditions, then measured multiple times to extract their distribution. Vset distribution for the cells centered on **μ = 1 V** have standard variation **σstd = 0.084**; for the cells centered on **μ = 2.5 V**, the standard variation is **σstd = 0.158 V**. To simplify the analysis we are assuming that the standard variation **σstd** of each cell varies linearly with **Vset**, fitting the two above experimental data points. **σstd** which is representing the stability of the **Vset** of a particular cell is 3–6 times smaller than the standard deviation σ of the entire population. As shown in Fig. 10.9, the wider we blank with "X" the population near the median, the lower the expected CRP error rate is. In this figure the challenges are derived from the experimental measurements on **Vset**, the responses are just examples.

In this statistical analysis we studied three cases for challenge generation: (1) blanking **±0.27 V** around the median value of **μ = 2.1 V**; (2) blanking **±0.54 V**, and (3) blanking **±0.81 V**. For response generation we are reducing the threshold value to **1.8 V** to sort out the "0"s from the "1"s. Such a reduction of the threshold reduces the CRP error rates due to the physical asymmetry of the **Vset** which has lower **σstd** at lower value. The **±0.27 V** blanking is not large enough, CRP error rates are too high, in the 80,000 ppm range. When the blanking is large enough, the likelihood that a bit from a response population can flip from its state as

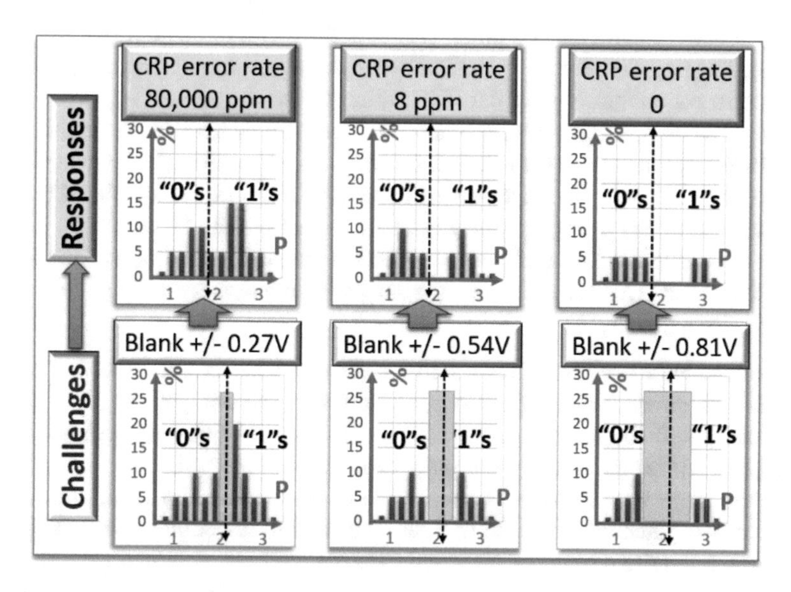

Fig. 10.9 Simulation CRP error (ReRam samples)

a challenge ("0" to "1" or the opposite) creating a CRP error is negligible: **both ±0.54 V** and **±0.81 V** blanking can yield CRP error rates below 8 ppm.

The impact of this error rate on the authentication cycle of a PUF stream of **N** bits can be calculated with the Poisson equation. If **P(n)** is the probability to have **n** failures over **N** bits, **p** is the probability to have one CRP mismatch due to errors, assuming **N = 128** and **p = 8 ppm** (case **±0.54 V**). **P(n)** is given by:

$$\mathbf{P(n) = \lambda^n / n!\, e^{-\lambda}} \tag{10.13}$$

$$\mathbf{\lambda = pN} \tag{10.14}$$

$$\mathbf{P(0) = 99.2\,\%;\ P(1) = 0.794\,\%;\ P(2) = 0.003\,\%;\ P(3) \approx 0.} \tag{10.15}$$

With these results, the probability that at least 126 bits over 128 CRP candidates are matching during an authentication cycle is almost certain. Such CRP stream error rates are lower than 98 % which are much below the generally accepted rate of 90 % for PUF authentications. The **±0.54 V** blanking correspond to **±1σ** of the **Vset** distribution, about 68 % of the cells are then blanked "X," 32 % are used for CRP generation. In this example, in order to generate a PUF of 128 bits, the size of the memory array needs to be in the 400 bit range.

Non-obviousness: The method based on ternary states has potential to enhance the non-obviousness, and unclonability of the PUF for the following reasons:

- The mapping of the "X"s can be encrypted and stored in the secure memory. An hacker trying to extract a response would have the difficult task to uncover the location within the memory array used for challenge generation.
- The secure element may communicate with the server through binary streams of data, keeping the ternary logic internaly. A hacker would then needs ternary logic to communicate with the PUF within the secure element.
- One way to manage ternary states in a secure element is to duplicate the information: manage a **(0, 1)** pair for a "0," a **(1, 0)** pair for a "1," and a **(1, 1)** or a **(0, 0)** for a "X.". Such an internal ternary bit structure can confuse DPA side channel analysis trying to differentiate the "0"s from the "1"s.
- There is little need to apply error correction methods in addition of the ternary state method. This can accelerate the authentication process leaving less time for side channel analysis.

Random Number Generators (RNG): As presented in Sect. 10.2.1 RNG are important to create hashing functions, and enhance the encryption. RNG that are generated purely by mathematical methods are by definition not really random when a hacker learns to apply the same mathematical method.

As shown in Fig. 10.10, it is possible to exploit the ternary method presented above to strengthen the RNG. In this case we want to select only the cells that are close to the threshold between "0"s and "1"s, the "X" states. When subject to one particular reading the stream of binary bits generated by testing the cells with "X"

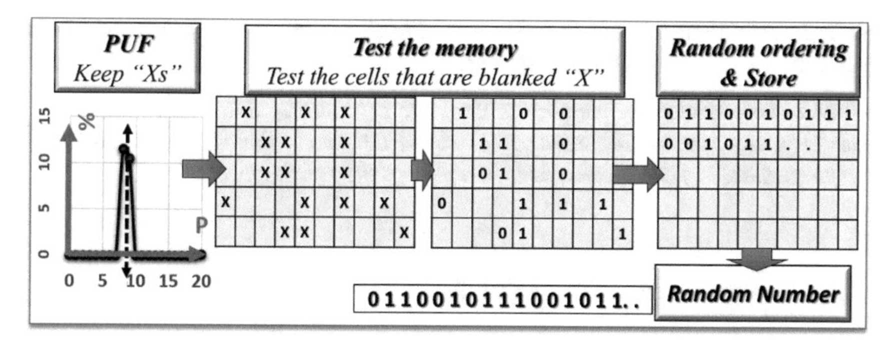

Fig. 10.10 Random number generation with ternary states

states will have a high level of randomness. For example if the cells with **Vset** that are **±0.1 V** from the median **μ = 2.1 V** are used for RNG, the likelihood that they will flip on the opposite bit after multiple reading is 43 %. Let us assume that a random number of 128 bits was generated with half of its bit having a probability **p = 0.43** to be a "0," and the other half with **p = 0.57**. Assuming binomial distribution, the likelihood that half of this population is at "0" is not statistically different than a population with **p = 1** for the first half and **p = 0** for the second half. Mathematical methods to randomly re-order these 128 bits will yield a higher level of randomness when each bit has a probability to flip back and forward close to **p = 0.5**. The usage of a population with **p = 0.43** is indeed much better than re-ordering a memory array filled with "0"s and "1"s. There are ways to further improve this randomness: add noise during the testing of the "0"s and the "1"s, or measuring the **Vset** at higher sweeping rates.

Implementing ternary states with SE: A first way to implement ternary-based PUFs on secure elements is to store the position of the blanked cells "X" somewhere in the secure memory. Then, during response generation, the state machine of the memory or memory management unit (MMU) can only exploit non-blanked cells. To enhance security the location of the blanked cells can be encrypted.

A second implementation is also based on the use of the memory array as is, while creating ternary states within the array itself. The memory array shall be segmented by pairs of cells to create the ternary states. The first of the two cells, called "active cell" shall be used to generate a PUF challenge, either a "0" or a "1" based on its parameter **P**. The second cell, called "companion cell," is to be used to differentiate the state of the active cell: for a state "0," a "0" is stored in this active cell, and a "1" is stored on the companion cell. Conversely for a state "1," a "1" is stored in the active cell, and a "0" in the companion cell. To store a blanked "X" state, the same bit shall be stored both in the active cell, and its companion cell. The MMU will then drive for the SE pairs of bits, (**0, 1**), (**1, 0**), and (**1, 1**) or (**0, 0**), while the external communication with the server is binary. During authentication, the responses are generated by testing only the active cells that were not blanked during the challenge generation, then challenge-response pairs are matched bit by bit. In both implementations there is no need to change the memory array within the SE, only internal logic changes are required.

What to remember: *the method to blank the cells susceptible to flip between 0"s and "1"s can reduce CRP error rates in a PUF, and enhance the unclonability. A "by product" of the method is to strengthen the randomness of RNGs.*

10.5 Usage of Machine Learning

Machine learning is an architecture that is used in cybersecurity to improve authentication, and detect abnormal behavior of the user, or the server, [36, 39]. On the terminal side, machine learning has been used for biometric authentication to track drifts of the user entries over time. Tara Seal from Info-security Magazine reported in September 2014, that the security company CA Technologies is using machine learning to combat credit card fraud. On the infrastructure side, machine learning has been widely used for computer security, and to protect web authentication. As shown in Fig. 10.11, Machine Learning Engines (MLEs) can be inserted into secure elements between the server, and the secure memory, to operate in a closed loop, without external intervention to avoid disclosing additional information during attacks. The MLEs can be then dedicated to the authentication process. The MLE can be the engine to track behavior of both the user, and the server to flag abnormal situations such as repetitive negative attempts to authenticate, or malicious changes in the power supply. The MLE can also enhance the effectiveness of the PUF, this is the subject of the study presented below.

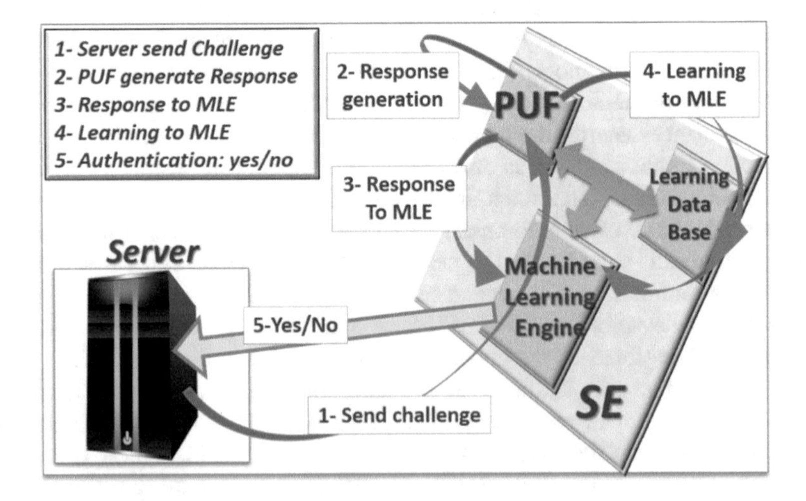

Fig. 10.11 PUF authentication with machine learning

Track the physical variations of the PUFs with MLE: The PUF responses and challenges are generated the same way with a PUF memory, and this as often as there is need for a fresh authentication, however the responses can vary overtime due to the natural variations of parameter P which underlie the PUF. This can create large CRP errors, and weaken the usefulness of PUFs.

A method to correct such a problem is described below step by step (see Fig. 10.12). In this method an MLE architecture is combined with a multi-state CRP generation process to track the drift of the responses, and determine if the drift is natural, or has changed due to a malicious entry.

- *Challenges*. As presented Sect. 10.3.2, the PUF Challenges shall be generated with the memory, "0s" are programmed in the cells where parameter **P** is below threshold, and a "1" is programmed above the threshold.
- *Responses*. The responses shall be generated with the same method than the challenges. Concurrently the cells shall be organized in n = **8** multiple states **i** by sorting out the value of parameter **P** for these cells. For example, when the response has N = 128 bits, the 16 cells with the lowest value are given the state **0**, the following 16 cells with the state **1**, all the way to the 16 cells with the highest value that are given the state **7**. That way the 128 bits of the PUF responses are sorted in 8 different states.
- *Vector of error—authentication **j***. Each state $\boldsymbol{i}$ has n_i cells such that $\sum_{i=0}^{i=n-1} n_{i-1} = \mathbf{N}$. For a given cell $\boldsymbol{k}$ that is part of the PUF, the CRP error between the Challenge C_k and the Response R_k is given by the Hamming distance $\boldsymbol{\Delta} CRP_k = |R_k - C_k|$. For the state $\boldsymbol{i}$ and its n_i cells, the average CRP error rate $\mathbf{E_i}$ is given by Eq. 10.16. The average error rate $\mathbf{E_0}$ to $\mathbf{E_{n\text{-}1}}$ for the $\boldsymbol{n}$ states and the authentication **j** is giving a Vector of Error $\mathbf{VE_j}$, Eq. 10.17:

$$E_i = \frac{1}{n_i} \sum_{k=1}^{k=n_i} |R_k - C_k|^{\infty} \quad (10.16)$$

$$\mathbf{VE_j} = (E_0,\ E_1,\ E_i, ., E_{n-1})_{\mathbf{j}} \quad (10.17)$$

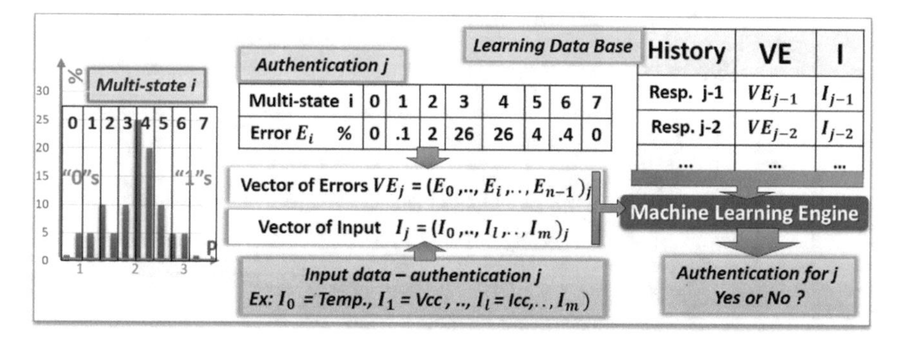

Fig. 10.12 CRP correction with machine learning

- *Vector of input—authentication* ***j***. The Vectors of Input are resulting from the measurements of environmental parameters surrounding the PUF such as ***Io*** temperature, $\boldsymbol{I_1}$ voltage bias conditions, $\boldsymbol{I_i}$ Rcurrent, and others:

$$I_j = (I_0, I_1, \ldots, I_i, \ldots, I_{n-1})_{\mathbf{j}} \tag{10.18}$$

- *Learning data base*. Considering that parameter **P** is a physical parameter, it is possible to develop a predictive model on how the responses are anticipated to vary when subject to drifts in the environmental parameters ***I***. This can be used to develop a learning data base which keeps track of **VE** when parameter **P** is subject to drifts. The learning database can also contain the history of previous authentications, $\mathbf{VE_h}$ and $\boldsymbol{I_h}$ with $\mathbf{h < j}$.
- *Final authentication* ***j***. To produce a "yes" or "no" for authentication j the MLE analyzes the vector of error $\mathbf{VE_j}$ which contain the average CRP error rates by state together with $\boldsymbol{I_j}$, and the learning data base. This exploits the predictability of parameters **P** when subject to environmental variations. On the contrary, a challenge brought by a malicious third party when matched with a fresh response would create a **VE** that is not consistent with the learning data base, thereby resulting in a negative authentication.
- *Experimental validation*. In this section we are giving a real example, based on ReRAM samples produced at Virginia Tech. This should help the reader to better understand the method presented above on how the MLE can analyze CRP errors, and make a determination on authentication **j**. A summary of the analysis is shown Fig. 10.13: At room temperature (20 °C) the ReRAM samples produced at Virginia Tech have a **Vset** median distribution $\boldsymbol{\mu} = \mathbf{2.1\ V}$ and standard variation $\boldsymbol{\sigma} = \mathbf{0.54\ V}$. If we assume that both the challenges, and the responses are generated at 20 °C, and the CRP errors are only created by the natural variations of the measurement of each cells. Assuming a binomial distribution as described Sect. 10.3.2 the expected vector of error $\mathbf{VE_{20\text{-}20}}$ is

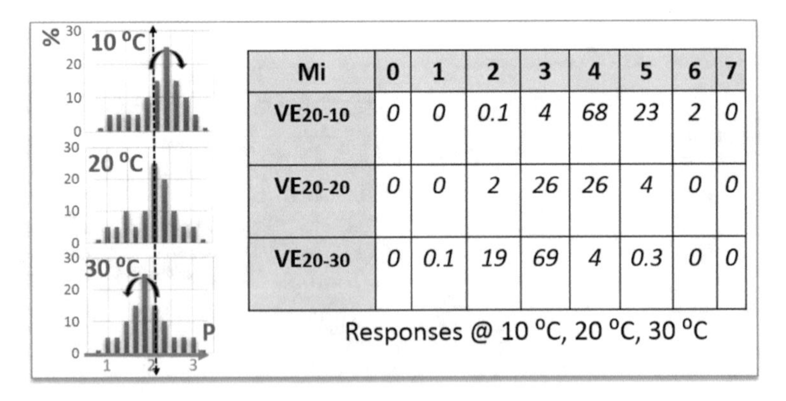

Mi	0	1	2	3	4	5	6	7
VE20-10	0	0	0.1	4	68	23	2	0
VE20-20	0	0	2	26	26	4	0	0
VE20-30	0	0.1	19	69	4	0.3	0	0

Fig. 10.13 Simulated CRP error rate (ReRAM, Vset)

$$\mathbf{VE_{20-20}} = (\mathbf{0},\ \mathbf{0.1},\ \mathbf{2},\ \mathbf{26},\ \mathbf{26},\ \mathbf{4},\ \mathbf{0.4},\ \mathbf{0}) \tag{10.19}$$

When the ambient temperature is reduced to 10°C, the distribution is shifting upward with a mean $\boldsymbol{\mu} = \mathbf{2.25\ V}$. If we assume that the challenge was extracted at 20 °C, and the response at 10 °C, the vector of error $\mathbf{VE_{20\text{-}10}}$ is

$$\mathbf{VE_{20-10}} = (\mathbf{0},\ \mathbf{0},\ \mathbf{0.1},\ \mathbf{4},\ \mathbf{68},\ \mathbf{23},\ \mathbf{2},\ \mathbf{0}) \tag{10.20}$$

Conversely if the ambient temperature is increased to 30°C, the distribution is shifting downward with a mean $\boldsymbol{\mu} = \mathbf{1.95\ V}$. If the challenge was extracted at 20 °C, and the response at 30 °C, the vector of error $\mathbf{VE_{20\text{-}30}}$ is

$$\mathbf{VE_{20-30}} = (\mathbf{0},\ \mathbf{0.1},\ \mathbf{19},\ \mathbf{69},\ \mathbf{4},\ \mathbf{0.3},\ \mathbf{0},\ \mathbf{0}) \tag{10.21}$$

The algorithm of the MLE can be set in two different ways: (i) the vector of error at various temperatures can be stored in the database of learning, so such error rate variation with temperature will be considered normal, or (ii) assuming that the temperature is measured concurrently with the response, the MLE can correct the anticipated drift resulting in lower CRP error rates. Both cases are not computing power intensive, and does not require large learning databases. For example, the storage of 1,000 different vectors in the learning database will occupy less than 1Kbyte of the embedded memory.

What to remember: *Machine learning can enhance the authentication process of the SE, and track hostile behavior. When combined with multi-state sorting, MLE can track the natural drift of the PUFs, reduce FRR, and FAR.*

10.6 Additional Enhancements

We are suggesting the following methods to further enhance SE:

Programmable and random active shielding: The aim of active shielding is to recognize physical attacks, and trigger an alteration of the SE with partial erasing of the stored information. Protective structures are inserted within existing and additional top-metal interconnection layers to shield the IC in case of physical attacks [40]. Signal layers are set to electrically detect attempts to probe or force internal modules in the IC. The randomness in the shape of the metal shielding is intended to make it hard to recognize. Signal detection methods that are programed to constantly vary can prevent some systematic techniques of attack.

Giant key authentication: Multi-factor authentications are often done sequentially, each factor separately. Hackers can then concentrate on the first factor to extract the first key, then move to the next factor. A way to enhance security is to create a giant authentication key combining multiple cryptographic keys such as passwords, pin codes, and reference keys. The authentication can be done thereby at once with all keys brought together to re-create the giant key. Facing a negative authentication the hacker will not be able to extract the content of one particular key. A simple method to create a combined key is to XOR the first key with the second key [41]. A second method called "edit distance" [9] is to form a giant key GK by inserting on the first key additional bits based on the distance defined by the second key. For example if the first key is a stream of binary bits, and the second key is a digital pin code, additional "0" are inserted as explained below:

$$\textbf{Key 1} = (1,\ 1,\ 0,\ 0,\ 1,\ 1,\ 0,\ 1,\ 0);\quad \textbf{Key 2} \text{ is pin } \textbf{23} \tag{10.22}$$

The resulting giant key is

$$\textbf{GK} = (1,\ 1,\ \underline{\mathbf{0}},\ 0,\ 0,\ 1,\ \underline{\mathbf{0}},,\ 1,\ 0,\ 1,\ 0) \tag{10.23}$$

The first digit "2" of the pin code is adding an extra "0" after the distance 2, and the second digit "3" is adding an extra "0" after the distance 3. Such giant key management can be done internally within the SE.

Chip design for security: Precautions need to be taken during the design of the SE, [42], however logic design engineers are not necessarily familiar with specific security requirements and could make mistakes. Design tool makers such as Tortuga Logic [43] developed back end tools that insert these predefined requirement in the database that generate the masks of the SE. Examples are

- Verify that the encryption/decryption protocols are enforced. The data transmitted along the buses should be encrypted/decrypted. All data and addresses transmitted to and from the chip, and within the chip that are relevant to security should be guaranteed to be encrypted.
- Verify that anti DPA measures are enforced. Implement current masking methods to scramble current consumption including performing dummy access operations to all memory modules, cache, ROM, NVRAM, and CAM. The current consumption of the actual program flow should be hidden. Leverage RNG and random wait states to further confuse DPA.
- Anti-tampering measures through the memory management unit. To spread and scatter the storage of the important data, and the cryptographic keys, all over the memory space. Use of error correction, and check sum methods to hide stored data. True hardware firewalls within the memory space of the critical elements such as embedded operating system, and critical applets. The use of cyclical redundancy checks (CRC) to verify data integrity, check errors, and follow ISO/IEC 7816 & 14443 standards.

Use of sensors: As discussed Sect. 10.4 relates to MLEs, we suggest the use of environmental sensors to capture ambient temperature and biasing conditions. Enhancing security of the SE can include light sensors on the IC to detect unwelcomed opening of the package, frequency sensors to check the internal clock, sensors and filters to monitor the external clock, external voltage sensors to check Vcc, and internal glitch and voltage/current sensors to detect an attack.

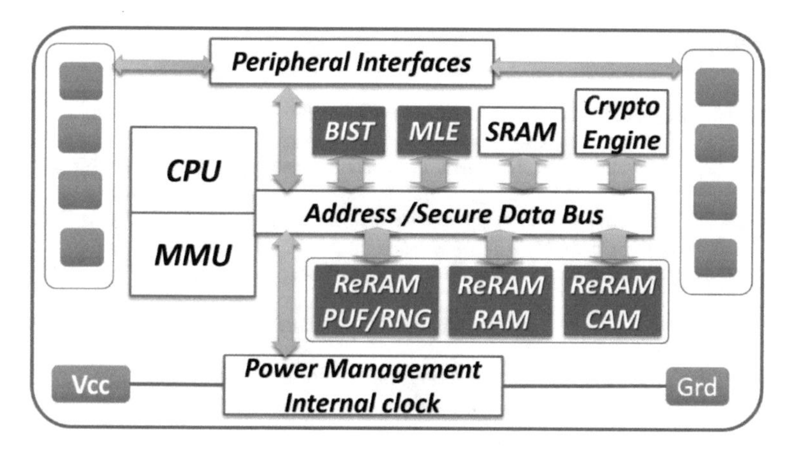

Fig. 10.14 Enhanced secure element with ReRAM

10.7 Summary

Figure 10.14 follows on Fig. 10.1 that incorporate the suggested methods to strengthen SE compatible with current ISO standard:

Replacement of Flash by ReRAM: The advantages include low power of operation for read/program/erase, and high performance.

Addition of Content Addressable Memory: CAMs allow the direct matching of stored passwords, pin codes, and other keys without exposing the keys.

Use of the embedded memory to generate PUFs: Embedded memories can be exploited to generate PUFs that are protected by the crypto-processor of the SEs. The communication to and from the server can then be encrypted.

Use of ternary logic: to blank the marginal cells and BIST integrated in the SE and reduce CRP errors, lower both FRR, and FAR. The natural randomness of these PUFs can also be exploited for RNG.

Use of MLE: MLE can strengthen the authentication processes, and generate stronger PUFs, differentiating the natural drifts from hostile random behaviors.

Various: Active shielding to prevent physical attacks, multi-function authentication with giant encrypted keys, build security upfront through chip design, sensors to detect side channel analysis.

References

1. Smartcard & security basics, www.smartcardbasics.com, www.cardlogix.com, sales@cardlogix.com, 2009 CardLogix Corporation
2. C. Medich, S. Swaminathan, K. Urban, S. Narendra, Maturity of smart card chip technology and its application to web security, in *Smartcard Alliance, Webcrypto 2014*
3. Global Platform, device technology, secure element access control, *Version 0.10.0; March 2012, Document Reference: GPD_SPE_013*
4. Card Payments Roadmap in the United States, *A Smart Card Alliance*, Feb 2011, Number PC-1100
5. Using Smart Cards for Secure Physical Access, *a Smart Card Alliance Report, Publication Number: ID-03003*, Jul 2013
6. H.X. Mel, D. Baker, *Cryptography Decrypted* (Addison-Wesley, 2000)
7. C.P. Pfleeger, S.L. Pfleeger, J. Margulies, *Security in Computing*, 5th edn. (Library of Congress, Person Education, 2015)
8. T. Cooijmans, J. de Ruiter, E. Poll, Analysis of secure key storage solutions on Android (ACM, 2011). ISSN 978-1-4503-3155
9. B. Cambou, Multi-factor authentication using a combined secure pattern, US Patent Application No 22938751, 16 Jul 2015
10. C. Krutzik, Solid state drive physical unclonable function erase verification device and method, US Patent Publication US 2015/0007337 A1
11. D. Merli, F. Stumpf, G. Sigl, Protecting PUF Error Correction by Codeword Masking; *IACR Cryptography, e-print archive 2013*, p. 334
12. R. Bez, E. Camerlenghi, A. Modelli, A. Visconti, Introduction to flash memory, in *Proceedings of the IEEE*, vol. 91, no. 4 (2003)
13. L. Crippa, R. Micheloni, I. Motta, M. Sangalli, Nonvolatile memories: NOR vs. NAND, in ed. by R. Micheloni et al. *Memories in Wireless Systems* (Springer, Berlin, 2008)
14. P. Kocher, J. Jaffe, B. Jun, Differential power analysis, in Crypto 99, LNCS 16666 (Springer, Heidelberg, 1999), pp. 388–397
15. G. Ghosh, M. Orlowski, Write and erase threshold voltage interdepèndence in resistive switching memory cells. IEEE Trans. Electron Dev. **62**(9), 2850–2857 (2015)
16. A. Makarov, V. Sverdlov, S. Selberherr, Modeling of the SET and RESET bipolar resistive oxide-based memory using Monte Carlo simulations, in *NMA 2010*. LNCS 6046 (Springer, Berlin, 2011), pp. 87–94
17. J.S. Meena, S.M. Sze, U. Chand, T.-Y. Tseng, Overview of emerging NVM technologies. Nanoscale Res. Lett. **9**, 526 (2014)
18. T.M. Maffit et al., Design considerations for MRAM. IBM J. Res. Dev. **50**(1) (2006)
19. B. Cambou, N. Burger, M. El Baraji, Apparatus system, and method for matching patterns with an ultra-fast check engine, US Patent No 8,717,794B2 (2014)
20. B. Cambou, Memory circuits using a blocking state, US 0atent Application No: 22728483, 24 Jun 2015
21. B. Cambou, ReRAM architectures for secure systems, US Application No 62/169957, 2 Jun 2015
22. Y. Jin, Introduction to hardware security, Electronics **4**, 763–784 (2015). doi:10.3390/electronics4040763
23. Z. Gong, M.X. Makkes, Hardware trojan side-channels based on PUF, in *Information Security*, vol. 6633. Notes in Computer Science (2011), pp 294–303
24. D. Naccache, Patrice. Frémanteau, Aug. 1992, Unforgeable identification device, identification device reader and method of identification, Patent US 5434917
25. R. Pappu, B. Recht, J. Taylor, N. Gershenfield, Physical one-way functions. Science **297** (5589), 2026–2030, 20 Sept 2002
26. R. Maes, P. Tuyls, I. Verbauwhede, A soft decision helper data algorithm for SRAM PUFs, in *IEEE International Symposium on Information Theory* (2009)

27. M. Hiller et al., Breaking through fixed PUF block limitations with DSC and convolutional codes, in *TrustED'13* (2013)
28. P. Prabhu, A. Akel, L.M. Grupp, W.K.S. Yu, G.E. Suh, E. Kan, S. Swanson, Extracting device fingerprints from flash memory by exploiting physical variations, in *4th International Conference on Trust and Trustworthy Computing* (2011)
29. D.E. Holcomb, W.P. Burleson, K. Fu, Power-up SRAM state as an identifying fingerprint and source of TRN, IEEE Trans. Comput. **57**(11) (2008)
30. T.A. Christensen, J.E. Sheets II, Implementing PUF utilizing EDRAM memory cell capacitance variation, Patent No.: US 8,300,450 B2 (2012)
31. X. Zhu et al., Daha Fazla, PUFs based on resistivity of MRAM magnetic tunnel junctions, Patents. US 2015/0071432 A1
32. E.I. Vatajelu, G.D. Natale, M. Barbareschi, L. Torres, M. Indaco, P. Prinetto, STT-MRAM-based PUF architecture exploiting magnetic tunnel junction fabrication-induced variability, ACM Trans. (2015)
33. A. Chen, Comprehensive assessment of RRAM-based PUF for hardware security applications (2015). ISBN *978-1-4673-9894-7/15/IEDM IEEE*
34. A. Gupta, Implementing generic BIST for testing kilo-bit memories, Master Thesis No-6030402 Deemed University Patiala India (2005)
35. D. Yamamoto, K. Sakiyama, K. Ohto, M. Itoh, Uniqueness enhancement of PUF responses based on the locations of random outputting RS latches, in *CHES 2011*, vol. 6917. Computer Science, pp 390–406
36. A. Joseph, P. Laskov, F. Roli, D. Tygar, B. Nelson, Machine learning methods for computer security, in *Manifesto from Dagstuhl Perspective Workshop 12371* (2012)
37. A. Casini, *Understanding Machine Learning Effectiveness to Protect WEB Authentication* (Universita Ca Foscari, Venezia, 2014)
38. S.Y. Kung, M.W. Mak, S. Lin, *Biometric Authentication: A Machine Learning Approach*. Information, System Science Series (Prentice Hall, 2004)
39. Y.V. Kaganov, Machine learning methods in authentication problems using password keystroke dynamics. Comput. Math. Model. **26**(3), 398–407 (2015)
40. S. Briais, J.M. Cioranesco, J.L. Danger, S. Guilley, D. Naccache, T. Porteboeuf, Random active shield, hal-0072569v2 (2012)
41. M. Robinton, S.B. Guthery, Efficient two-factor authentication, US Patent Application US 2010/0235900 A1 (2010)
42. What Makes a Smart Card Secure? *A Smart Card Alliance, White Paper, Publication Date: October 2008, Publication Number: CPMC-08002*
43. Tortuga logic web site, www.tortugalogic.com

Index

Note: Page numbers followed by f and t indicate figures and tables respectively

L. Bossuet and L. Torres (eds.), *Foundations of Hardware IP Protection*,
DOI 10.1007/978-3-319-50380-6

Printed in the United States
By Bookmasters